APPLIED DATA ANALYSIS FOR URBAN PLANNING AND MANAGEMENT

Sara Miller McCune founded SAGE Publishing in 1965 to support the dissemination of usable knowledge and educate a global community. SAGE publishes more than 1000 journals and over 800 new books each year, spanning a wide range of subject areas. Our growing selection of library products includes archives, data, case studies and video. SAGE remains majority owned by our founder and after her lifetime will become owned by a charitable trust that secures the company's continued independence.

Los Angeles | London | New Delhi | Singapore | Washington DC | Melbourne

APPLIED DATA ANALYSIS FOR URBAN PLANNING AND MANAGEMENT

Alasdair Rae
Cecilia Wong

SAGE

Los Angeles | London | New Delhi
Singapore | Washington DC | Melbourne

SAGE

Los Angeles | London | New Delhi
Singapore | Washington DC | Melbourne

SAGE Publications Ltd
1 Oliver's Yard
55 City Road
London EC1Y 1SP

SAGE Publications Inc.
2455 Teller Road
Thousand Oaks, California 91320

SAGE Publications India Pvt Ltd
B 1/I 1 Mohan Cooperative Industrial Area
Mathura Road
New Delhi 110 044

SAGE Publications Asia-Pacific Pte Ltd
3 Church Street
#10-04 Samsung Hub
Singapore 049483

Editor: Alysha Owen
Assistant editor: Charlotte Bush
Production editor: Martin Fox
Copyeditor: Catja Pafort
Proofreader: Rebecca Storr
Indexer: Silvia Benvenuto
Cover design: Shaun Mercier
Typeset by: C&M Digitals (P) Ltd, Chennai, India
Printed in the UK

Library of Congress Control Number: 2020950419

British Library Cataloguing in Publication data

A catalogue record for this book is available from the British Library

ISBN 978-1-5264-9700-0
ISBN 978-1-5264-9699-7 (pbk)

CONTENTS

ABOUT THE EDITORS

Alasdair Rae is the founder of Automatic Knowledge Ltd, a UK-based data and insights company, focusing on spatial data analysis and the built environment. Prior to that, he was a Professor of Urban Studies and Planning at the University of Sheffield. He is a Fellow of the Royal Scottish Geographical Society, a recipient of the Royal Town Planning Institute's Sir Peter Hall Award for Wider Engagement, a former Commissioner of the UK2070 Commission, and a winner of the Royal Statistical Society's 'Stat of the Year'. His most recent academic work has focused on spatial analysis, deprivation, housing markets and megaregions, and his work frequently appears in the national and international media. He has a PhD from the University of Liverpool, an MA from The Ohio State University and a BA from the University of Strathclyde.

Cecilia Wong is Professor of Spatial Planning and Director of Spatial Policy and Analysis Lab of the Manchester Urban Institute, the University of Manchester. She is Fellow of the UK Academy of Social Sciences; Fellow of the Royal Town Planning Institute; and Chair of the UK 2021 Research Excellence Framework 'Architecture, Built Environment and Planning' sub-panel. She has research expertise in strategic spatial planning, policy monitoring and analysis, urban and regional development, and housing and infrastructure planning, and urban planning in China. She was a commissioner of the UK2070 Commission on Regional Inequalities, a commissioner of the Labour Party's Lyons

Independent Housing Review, a member of Department for Communities and Local Government's expert panel on housing and planning, an expert panel member of the European Commission's Urban Audit II, and an expert group member of the UN-Habitat City Prosperity Index.

ABOUT THE CONTRIBUTORS

Benjamin Bechtel holds a professorship in Urban Climatology at the Ruhr-University Bochum, Germany. He serves as board member of the International Association for Urban Climate, Associate Editor for *Frontiers in Remote Sensing* and as a reviewer and guest editor for several international journals.

Yiqiao Chen is a PhD Candidate in the Department of Land Economy, University of Cambridge. Yiqiao's research focuses on utilisation of multi-sourced data in the big data era, and integration of complexity theory, urban data analytics and smart city governance, with a special focus on smart transport.

Jason Ching has had joint appointments as a Meteorologist/Physical Scientist with the US National Oceanographic and Atmospheric Administration (NOAA) and the US Environmental Protection Agency (EPA) from 1970 to 2010; he led the development of the EPA's prototypic Community Multiscale Air Quality (CMAQ) modelling system. Dr Ching is currently a senior research scientist at University of North Carolina Institute for the Environment in Chapel Hill, NC, and is a lead for the WUDAPT (World Urban Database and Access Portal Tool), an IAUC Initiative (2012–present).

Oliver Dawkins is a researcher on the Connected Environments project at the Bartlett Centre for Advanced Spatial Analysis (CASA), University College London, and formerly worked on the Building City Dashboards project at Maynooth University. His research focuses on the use of geographic information in 3D, sensors and real-time data, mixed realities and advanced visualisation technologies.

Jimmy Chi Hung Fung is a professor in the Department of Mathematics of the Hong Kong University of Science and Technology. Fung's speciality is to apply advanced atmospheric and chemistry transportation models to study atmospheric science and air quality over Hong Kong and Pearl River Delta regions down to street scales.

Rob Kitchin is a professor in Maynooth University Social Sciences Institute and Department of Geography. He was a European Research Council Advanced Investigator on the Programmable City project (2013–2018) and a principal investigator on the Building City Dashboards project (2016–2020) and has published widely on the relationship between technology, space and society.

Heeseo Rain Kwon is a Post-Doctoral Research Assistant in the Department of Real Estate and Planning, University of Reading. Rain holds a PhD in Land Economy from the University of Cambridge and is currently a PDRA for a Medical Research Council UKPRP project on healthier urban development. Her research interests focus on the application of behavioural theories in urban development and planning, dynamic simulation using agent-based modelling (ABM) and GIS-based spatial analysis.

Lun Liu is an Assistant Professor & Research Scientist in the School of Government, Peking University. Dr Liu holds a PhD in Land Economy from the University of Cambridge. Her research focuses on the application of data science in urban management and urban policy analysis.

Gerald Mills is a physical geographer in the School of Geography at UCD. His research focuses on the climates of cities and he has served as President of the Geographical Society of Ireland and of the International Association for Urban Climates. He is co-author of *Urban Climates* (CUP) and of *The Urban Heat Island: A Guidebook* (Elsevier).

Edward Ng is the Yao Ling Sun Endowed Chair in Architecture in the School of Architecture of The Chinese University of Hong Kong. He specialises in Green Building, Environmental and Sustainable Design, and Urban Climatology for City Planning.

Haifeng Niu is a PhD Candidate in the Department of Land Economy, University of Cambridge. Haifeng's research focuses on how the intersection of data science and advanced spatial analysis contributes to better urban policy-making and smart cities management. His research is centred in urban big data (particularly social media Data, POI data) mining, spatial data science & AI, geovisualisation and planning support systems.

Jose Reis is Principal Policy Officer – Economic Strategy, Greater London Authority. Dr Reis is also an Affiliated Lecturer at the Department of Land Economy, and a Visiting Researcher at LISA Lab, Department of Land Economy, University of Cambridge. His research interests link spatial analysis and public policy; focusing on urban and regional policy, and in helping cities and regions to prosper. Dr Reis has expertise in economic development, urban planning and regeneration, devolution, urban resilience and spatial analysis.

Chao Ren is an Associate Professor in the Faculty of Architecture at the University of Hong Kong. Her research interest is Sustainable Urban and Environmental Design and Urban Climatic Application in Urban Planning.

Linda See is a Senior Research Scholar in the Advancing Systems Analysis Program at the International Institute for Applied Systems Analysis (IIASA) in Austria. Her research explores the synergies between Earth Observation and citizen science related to land cover and land use.

Elisabete A. Silva is Professor of Spatial Planning in the Department of Land Economy, University of Cambridge. Professor Silva is the Director of the Lab of Interdisciplinary Spatial Analysis (LISA Lab) at the Department of Land Economy, University of Cambridge. Her research interests are centred on the application of new technologies to spatial planning in particular city-metropolitan dynamic modelling (CA, ABM, NN), and complexity and spatial analysis in a context of data science, big data and smart cities.

Duncan A. Smith is a lecturer in GIS and visualisation at the Centre for Advanced Spatial Analysis, University College London. His research interests include urban sustainability, transportation accessibility and interactive web visualisation.

Jason Wai Po Tse graduated with a Master of Philosophy degree in Atmospheric Environmental Science from Hong Kong University of Science and Technology. His research focuses are examining how to improve weather and climate prediction in urban areas and investigating the meteorological impact of urbanisation.

Lei Wang is Associate Professor in the Key Laboratory of Watershed Geographic Sciences, Nanjing Institute of Geography and Limnology, Chinese Academy of Sciences and Honorary Research Fellow in the Department of Planning and Environmental Management, the University of Manchester.

Zheng Wang is Lecturer at the Department of Urban Studies and Planning, the University of Sheffield. His research interests include the neighbourly relations and community engagement of rural migrants and the long-term social impacts of large-scale urban development projects. He has published in journals including *Antipode, Environment and Planning A, International Journal of Urban and Regional Research* and *Urban Studies*.

Pak Shing Yeung graduated with a Master of Philosophy degree in Environmental Science, Policy and Management from the Hong Kong University of Science and Technology (HKUST). His research focuses are investigating the meteorological impact of urbanisation and examining how to improve weather model prediction in urban areas.

Gareth W. Young is a postdoctoral research fellow on the V-SENSE project at Trinity College Dublin, and formerly worked on the Building City Dashboards project at the National Centre for Geocomputation (NCG), Maynooth University. Gareth's research focuses on evaluating real-world uses and design effectiveness of cross-reality (XR) technology using human–computer interaction (HCI) evaluation techniques.

Adipandang Yudono is Senior Lecturer in the Department of Urban and Regional Planning, Brawijaya University, Indonesia.

Tomasz Zawadzki is an Assistant Professor at ARUCAD – Arkin University of Creative Arts and Design in Cyprus, visiting scholar at PolyU in Hong Kong and formerly worked on the Building City Dashboards project at Maynooth University. Tomasz's research focuses on XR (VR/MR) in architecture and health care.

1

INTRODUCTION: THE PROMISES AND PITFALLS OF DATA FOR URBAN PLANNING

Alasdair Rae

Introduction

This is a book about analysis and planning in relation to the built environment. Written by authors from across the globe, it provides a new perspective on developments in data and analytics as they relate to urban planning and management. However, before we go any further, we need to be clear that it is not an idealistic, utopian thesis on the power of data and analytics to transform societies and make people's lives better. If the planning and technology literature tell us anything, it tells us that such notions are at best naïve and at worst, dangerous (e.g. Boeing et al., 2021).

Rather, what we seek to do with this text is explore a number of issues around data, analysis, planning and urban management in relation to examples from a range of countries. We look at different topics in each chapter, written by authors who come from different methodological traditions. In doing so, we hope that we can provide a balanced view of the subject and help situate the topic within its wider context. Yet, despite our caution about the risks associated with what might be termed 'big data hyperbole' (e.g. Rae and Singleton, 2015), recent developments in software, open data and online platforms have led to a proliferation of data-centric dashboards and platforms. This has in part been motivated by a desire among urban administrators to 'know' their cities better, to deliver 'better' services and to increase 'efficiency' for citizens. Such desires are laudable, but we wish to emphasise that questions relating to what can be 'known' or 'measured' very rarely have simple answers.

Therefore, we would urge readers to approach this text, and individual chapters, with the knowledge that the world we seek to understand is messy, complex, ill-defined and unpredictable. For older scholars, such a statement may bring to mind Rittel and Webber's (1973) notion of 'wicked problems'; that is, those problems of the social world where there are no 'optimal solutions'. The current COVID-19 crisis that has swept the globe as we write is a timely, troubling reminder of this.

Despite our caution and our critical mind-sets as a collective group of writers whose work focuses on data, analysis and the built environment we also recognise that data and technology can be a powerful force for good. If 'good' outcomes are to emerge from new technologies and methods, then they must be seen not as weapons to be wielded, but as tools of empowerment. Examples of this can be seen throughout the book, including in Chapter 4, where Adipandang Yudono looks at the crowdsourcing of geographic information for planning and management in Indonesia. This bottom-up approach helps demonstrate the enabling and democratising potential of new technologies.

The rest of this introduction is structured as follows. The next section considers what is new in analysis and planning, but from a zoomed-out perspective. This helps provide the wider context for the book, but also helps highlight some of the emerging critical literature on the topic. Following this, we provide a wider thematic overview of the text and its authors, with reference to individual chapters. In order to highlight the main ideas and findings of individual chapters, we then provide overview summaries as a way of helping readers turn to the topics of most relevance to them. The book can of course be read from cover to cover, and we encourage this, but it has also been designed as a reference text so that readers may dip in and out of individual chapters as required. Finally, the conclusion once again highlights the power and potential of the methods and tools we discuss, but urges readers to remain vigilant about potential pitfalls.

Analysis and planning: what's new?

At the time of writing (late 2020), the answer to the question of 'what's new in analysis and planning?' is the same as it is for almost every sector of society across the globe: COVID-19. Yet, like previous global crises, this too shall eventually pass and we believe there are likely to be new opportunities for growth and development in its aftermath. However, even before the emergence of the COVID-19 pandemic, global forces were acting to reshape our world.

One of the most obvious and significant of these is encapsulated in the term 'the Asian century'. This refers to the growing global importance, and predicted dominance, of Asian culture, technology and power as the 21st century progresses. Although the phrase emerged in the 1980s, it is only in recent years that it has begun to take hold as a way of understanding the world. Thus Kohli et al. (2011), in their *Asia 2050* book, discussed in detail this historic transformation and authors such as White (2011) have examined the role of Western nations in relation to the 'Asian century'. There are multiple ideas underlying this concept, but a major one is demographics. The population of China and India combined represents about 38 percent of the world's population in 2020 and, overall, 61 percent of the global population lives in Asia as a whole (United Nations, 2020). Furthermore, the population of sub-Saharan Africa is projected to double by 2050. Current global population settlement patterns, and the high population density of India and China in particular are illustrated in Figure 1.1, which helps emphasise the demographic dominance of Asia in particular.

Figure 1.1 A global population density map, highlighting the population of Asia

Of course, these demographic trends have been unfolding for decades, and on their own would not necessarily lead one to proclaim the emergence of the 'Asian century'. However, these demographic drivers have perhaps reached a kind of tipping point and, when combined with other factors they give rise to a new global context within which we must understand urban growth, development and analysis.

Combined with the emergence of new, more accessible technologies – and particularly open source software and data – we can highlight the following five important trends under the rubric of 'what's new', as below.

1. **Demographics:** as discussed above, the demographic balance of the world has shifted, slowly but surely, over the past four decades in particular and has given rise to the idea of the 'Asian century'. Combined with growth elsewhere (e.g. Latin America, sub-Saharan Africa) and demographic stagnation or decline in more advanced economies (e.g. Japan, Germany) this is a reminder that our analyses of cities, people and places does not take place in a static environment and that the concepts and ideas of specific times and places may no longer serve the purpose they once did.
2. **Technology and data:** the rise of 'tech' has of course been something of a double-edged sword, since it has the potential to both empower and disempower individuals. In the world of planning and urban management, however, the rise of open data created as part of the OpenStreetMap project, and of open source geographic and data tools such as QGIS and R, has lowered barriers to entry and allowed for genuine community-sourced contributions. Nonetheless, it is our view that most of the potential benefits of these developments are yet to be unleashed and that on their own they cannot disrupt long-established power structures.
3. **Place:** the importance of individual places remains high. Whether we consider the growth and development of global megacities over the past four decades (e.g. Hoole et al., 2019), or the urban hierarchies within a country, location matters. From a planning and urban management point of view, this is of course obvious yet it is not necessarily recognised at the national policy level. Hence, we see publications like the United Nation's *New Urban Agenda* (2016), which seek to centre the urban and embed challenges like inequality and poverty in national and local urban strategies.
4. **Politics and power:** we must also acknowledge that analysis, planning, urban management and the development of new places always takes place within specific political contexts. Thankfully, planning scholars in general have long recognised the necessity of understanding such themes (e.g. see Flyvbjerg's classic study of 'rationality and power', 1998) but this critical outlook has not always been as central when it comes to analytical and data-centric work. When we refer to 'politics' here, we mean national politics in terms of formal policies, but also the more informal, micro-politics of regions, cities and towns across the globe. Thus, who holds power and how it is exercised forms a critically important backdrop to any work in this field. The key point here is that when we are studying analysis and planning, we must understand them as being inextricably bound up in politics.
5. **Global economy:** any view of the ways in which analysis and planning operate in practice must also recognise the economic context. Just as global demographics have changed over the past four decades, so has the economic wealth of nations. Thus, we have the so-called

BRICs – the nations of Brazil, Russia, India and China. This term was coined by economist Jim O'Neill (2001) and has brought into focus the growing economic power of these emerging economies. The ways in which the rise of these economics has impacted upon societies globally is sometimes obvious (e.g. in relation to China's dominance in manufacturing), yet the key point here is that none of this happens in isolation. For example, the rise of manufacturing expertise in China has had very real implications for manufacturing, and the built environment in former manufacturing heartlands in Europe and the United States. The rise of the so-called 'MINT' economies of Mexico, Indonesia, Nigeria and Turkey provides further evidence that such trends are only likely to accelerate.

In answer to the 'what's new?' question in the title of this part of the chapter, in one sense the answer is 'everything'. That is, if we were writing this text in 1980 the urban, economic, political and demographic contours of the world would look very different to today. The addition of the COVID-19 pandemic adds another level of complexity and uncertainty. Unlike 1980, however, we now have enhanced levels of global connectivity, seamless global communications, almost limitless data, and the technology to analyse and make sense of it. These *are* new, and often give rise to a kind of unbridled optimism about the world and our ability to understand it. Yet, as recent literature on 'platform capitalism' or the rise of Airbnb has shown, such developments have real-world impacts, and they are never neutral or free of consequence (Ferreri and Sanyal, 2018; Leszczynski, 2020). A simpler way of saying this might be that *new* things are not always *good* things.

Themes we explore in this book

Following this *Introduction*, there are eight main substantive chapters in the book, followed by the *Conclusion*. In each of these chapters, the authors tackle different subjects, using a variety of case study examples. There is diversity both in terms of the origins of our authors, but also in relation to their specific areas of expertise.

In such an enterprise there is of course a risk that the individual parts can appear too disparate, yet we do not believe this is the case here. In fact, we sought from the start of this project to bring together a group of authors who would say different things, from different perspectives, on the trends currently unfolding in the broad area of analysis and planning as it relates to urban management across the world. Nonetheless, on first reading of the text, it may not be immediately obvious what the over-arching themes are, so we set them out here before going on to provide short introductions to each chapter.

Theme 1 – *Understanding complexity*. The first theme of the book is about the desire, or need, to understand complexity in urban planning and management in an inherently complex world. That is why Chapter 2 opens by discussing complexity right from the beginning. It also forms the backdrop for several other chapters, including Chapter 9. The question of whether we can ever truly 'know' or 'understand' the world through analysis and data is an ontological one, but we believe it is possible to understand the world better through the approaches and examples provided in this book.

Theme 2 – *The danger of seeing data as a panacea.* A 'panacea' is another word for a cure-all, or 'a solution or remedy for all difficulties or problems', to use a dictionary definition. We have discussed above how we do not view technology, data and analysis as a panacea, and this theme is an important element of the book. Chapter 3 provides some useful insights in this regard, both in relation to opportunities and challenges.

Theme 3 – *The potential of data, technology and analysis to 'make sense of the world'.* The relevance of this theme should be evident from a reading of any chapter in the book, but particularly so in Chapters 4, 7 and 8. The idea that we can make more sense of the world through such approaches is of course not new, but it is true that new methods and techniques represent a significant advance on what has been available to planners in the past.

Theme 4 – *The democratising potential of data.* Related to the previous theme is the notion of the democratising potential of data. The risk of hyperbole here is high, so we need to be clear that what we mean here is that the wider availability of data to a wider group of stakeholders than ever before has the potential to create a more level, more democratic playing field when it comes to decision-making. However, the key word in this theme is 'potential'. Yet the examples of Chapters 5 and 6 illustrate what is possible when data is made available to all in clear, compelling ways.

Theme 5 – *The power of data in tackling global challenges (e.g. climate change).* Among the many challenges that can be described as truly 'global' are inequality, sustainability, ageing and climate change. Many of these are embedded in the United Nation's Sustainable Development Goals, of which there are seventeen in total, from eliminating poverty to creating partnerships. Many of our chapters touch on this topic, but Chapter 7 in particular highlights the power of data analytics in relation to increasing understanding of climate change.

Theme 6 – *Data as a real-world resource.* Despite our caution around the idea of data as a panacea, we are keen to emphasise the utility of data as a real-world resource that can be a force for good. It is therefore important to have case studies which demonstrate this and in that regard Chapters 4 and 7 provide ample evidence of this.

Theme 7 – *Data does not have all the answers – or sometimes any of the answers.* The final theme of the book serves as a reminder that we need to remain circumspect when approaching the topic. There are many scenarios in urban planning and management where the answers are not to be found in data, either because it cannot capture the nuance and subtlety of the problem, or because it simply does not exist, or is too sparse. One example of this might relate to areas of the world for which there is no good quality geographic data for use in a GIS. Another example might relate to understanding processes of urbanisation and change in individual cities. That is why we are particularly pleased to include Chapter 9, which provides a fascinating case study of a data-sparse setting.

Chapter overviews

In *Chapter 2*, Elisabete A. Silva and colleagues ask 'what's new in urban data analytics?'. The chapter begins by highlighting the complex nature of urban problems and the need for evolution in the theoretical paradigms we use. The authors highlight the power and potential of 'big data', and also discuss the technical challenges associated with it. The need to properly integrate time and space into analyses of the urban is then discussed, before the authors move on to reflect on further considerations such as privacy and anonymisation. The chapter concludes with a salient reminder that 'we need to start to make sense of the legal, economic and political imperative in urban data'.

Complementing Silva et al.'s work, in *Chapter 3*, Cecilia Wong reflects on the opportunities and challenges associated with indicators and urban planning, in the context of what is currently talked about as 'big data'. In this chapter, Wong helpfully reminds us that many argue that 'big data' is not new at all, but more of a slippery concept. The concepts of 'volume', 'velocity' and 'variety' in relation to big data are highlighted, in addition to 'veracity', or truthfulness referred to in Chapter 6. Taking a step back from more in-vogue discussions about 'big data', the author here helpfully establishes the wider international context for urban discourse and the development of indicators and provides some seminal references, including Rae and Wong's 'four principles' from 2012. This chapter serves as a useful reminder that whatever the pitfalls might be, a data-centric approach is at the heart of urban planning and management and as such we must be pragmatic about how we engage with it.

Chapter 4 then takes a closer look at a specific case study, in which Adipandang Yudono explores the potential for crowdsourced geographic information to be used in urban planning and management in Indonesia. This is a more technical chapter, with a geospatial focus, and it highlights the growing use, and power, of GIS technology in the urban planning and management context. It also highlights the growing utility of crowdsourced geographic information, with particular reference to the OpenStreetMap (OSM) project. This has been of particular benefit in previously under-served or data-sparse nations where the geographic infrastructure is not as developed or well established as in wealthier nations or regions. It is in this context that the author highlights the usefulness of the Indonesia Humanitarian OpenStreetMap Team (HOT) and the mapping of large sections of the urban fabric in recent years. Yudono then introduces the idea of community empowerment and demonstrates that new geospatial technologies like OSM have an important role to play in democratisation.

In *Chapter 5*, Duncan A. Smith takes the reader through a series of design and narrative techniques for urban data visualisation, using his extensive catalogue of work in this field. Drawing upon key developments in the field, and situating his work within the wider context, Smith focuses on using data visualisation tools in public-facing applications. Combining technical and design-related ideas, a three-fold category of data visualisations is presented, comprising 'exploratory', 'narrative' and 'analytics' approaches. We can see examples of these elsewhere in the text (e.g. Chapters 4 and 6) so this is also a useful reference chapter for understanding how to position such work more broadly. A number of innovative examples are then presented, and these help demonstrate the limits of what is now possible with open data and geospatial technology.

Following on from Smith's tour of design and narrative techniques, and his categorisation of approaches, in *Chapter 6* Oliver Dawkins and colleagues discuss 'city dashboards' alongside 3D geospatial technologies for urban planning and management. The authors here at once combine the theoretical and the technical and show that new spatial media provide citizens and stakeholders with a very important way of making sense of complex data. This is particularly true in relation to empowering those without the skills to create their own. However, the authors are also careful to point out that such technologies, however deployed, are 'not without critique and shortcomings'. Thus, this chapter is a particularly important contribution if we wish to understand both the utility of new technologies in an urban context, as well as the main concerns about their use.

Chapter 7 takes a different approach, and looks at the linked issues of data analytics, urban form, and climate change. In this chapter, Chao Ren and colleagues make explicit reference to the United Nation's 17 Sustainable Development Goals (SDGs) and the *New Urban Agenda* mentioned above. They highlight the fact that, at present, a consistent database on global cites at scales suitable for scientific inquiry and policy formulation does not exist, and how this is a problem. Thus, this chapter can be thought of in relation to an assessment of problems and needs of urban data and analytics, in relation to the specific subject matter of climate change. The authors look at the World Urban Database and Access Portal Tools (WUDAPT) and use a number of different examples to help illustrate their utility, including the Pearl River Delta (PRD) megalopolis. This chapter provides an excellent demonstration of the ways in which agencies have tried to adapt to emerging challenges through coordinated approaches.

In *Chapter 8*, Lei Wang also provides a more focused case study example. In this chapter, the subject matter of data analytics and modelling is presented in the context of high-speed rail in China. This demonstrates the centrality of data and modelling in a contemporary context, in a nation which is rapidly expanding its high-speed rail capacity. Yet it also demonstrates the ways in which different data sources can be woven together (e.g. Table 8.2), to help support infrastructure-related decision-making. Whilst recognising the importance of such developments, Wang also considers which cities benefit and which do not – with reference to 'winners' and 'losers'. This is an important element because, as we have argued above, the rise of data and analytics in urban planning and management will always lead to uneven and unequal outcomes without careful oversight.

Chapter 9 then offers readers something different, but very important. In this chapter, Zheng Wang explores the issue of migrant–local relations in Shanghai, in a research setting characterised by a *lack*, rather than surplus, of data. As Editors, we felt it was important to remind readers that the data rich contexts many of us work within are not a universal phenomenon and in fact in many instances we have to consider different methodological approaches. Thus, in this chapter, Wang explains how a household survey has helped to investigate migrant–local relations in Shanghai in more depth, and in particular the nature of intergroup relations between rural migrants and native urban residents. This is very much an issue relevant to urban planning and management, since there are more than 245 million rural migrants living in Chinese cities, yet it cannot be measured or understood without a different approach to data collection and Wang's approach demonstrates how, in the absence of good quality official data, this research topic can be understood. The approach here is both qualitative and quantitative and demonstrates that when we are faced with a lack of data on an important topic, it is not necessarily a barrier to knowledge.

In *Chapter 10*, the final part of the book, Cecilia Wong looks beyond analytics to ask bigger questions about planning and urban management in general. An important conclusion is that different methods and techniques need to be deployed flexibly, in order to meet the needs of policy analysis under specific spatial contexts. When we look back on the various chapters outlined above, we can see this to be true, and particularly so in the case of the data-sparse setting in Wang's *Chapter 9*. Wong also points to the importance of understanding how technology and governance intersect, an interaction which speaks to the 'politics and power' trend outlined in this chapter. In addition, and despite all its successes, Wong highlights the 'conundrums of data science', pointing out that we are still very much in the experimental stage of these emerging analytical techniques, despite what the sometime hyperbolic literature might tell us. For this reason, we should see new analytical approaches as being complementary rather than displacing existing knowledge on planning and the built environment.

Conclusion

In this introductory chapter we have attempted to set the scene for what follows in the rest of the book. There are common themes and trends running through the text, yet at the same time each chapter offers something different and new. We hope readers will understand that, despite our caution about data-driven approaches, we also think there is much to be gained by taking an analytical approach to urban planning and management. Or, perhaps it is more appropriate to say that we believe such approaches should be incorporated and embedded in existing systems of knowledge, where it is appropriate to do so. This might mean the inclusion of a data dashboard for a city like Dublin as a way of providing new insights to policymakers and the public, or it might mean using community sourced geospatial data to improve map quality in Indonesian cities. We also hope readers approach the book with one eye on the wider context, particularly in relation to the five trends we highlight above: demographics, technology and data, place, politics and power, and the global economy. It is only by doing so that the context of this book can make sense. At the time of writing, in late 2020, the ongoing COVID-19 pandemic highlights the necessity of coordinated urban, regional and international thinking. We believe such an approach is also useful in urban planning and management.

References

Boeing, G., Besbris, M., Schachter, A. and Kuk, J. (2021) 'Housing search in the age of big data: Smarter cities or the same old blind spots?', *Housing Policy Debate*, 31 (1): 112–26.

Ferreri, M. and Sanyal, R. (2018) 'Platform economies and urban planning: Airbnb and regulated deregulation in London', *Urban Studies*, 55 (15): 3353–68.

Flyvbjerg, B. (1998) *Rationality and Power: Democracy in Practice*. Chicago, IL: University of Chicago Press.

Hoole, C. Hincks, S. and Rae, A. (2019) 'The contours of a new urban world? Megacity population growth and density since 1975', *Town Planning Review*, 90 (6): 653–79.

Kohli, H.S., Sharma, A. and Sood, A. (eds) (2011) *Asia 2050: Realizing the Asian Century*. New Delhi: SAGE.

Leszczynski, A. (2020) 'Glitchy vignettes of platform urbanism', *Environment and Planning D: Society and Space*, 38 (2): 189–208.

O'Neill, J. (2001) 'Building better global economic BRICs' (Global Economics Paper 66). London: Goldman Sachs. https://www.goldmansachs.com/insights/archive/archive-pdfs/build-better-brics.pdf (accessed December 14, 2020).

Rae, A. and Singleton, A. (2015) 'Putting big data in its place: Regional studies and regional science perspective', *Regional Studies, Regional Science*, 2 (1): 1–5. doi.org/10.1080/21681376.2014.990678

Rae, A. and Wong, C. (2012) 'Monitoring spatial planning policies: Towards an analytical, adaptive and spatial approach to a "wicked problem"', *Environment and Planning B: Planning and Design*, 39: 880–96.

Rittel, H.W.J. and Webber, M.M. (1973) 'Dilemmas in a general theory of planning', *Policy Sciences*, 4: 155–69. doi.org/10.1007/BF01405730

United Nations (2016) *New Urban Agenda* (Habitat III, Quito). United Nations Conference on Housing and Sustainable Urban Development. https://uploads.habitat3.org/hb3/NUA-English.pdf (accessed December 14, 2020).

United Nations (2020) 'Population'. Geneva: United Nations. https://www.un.org/en/sections/issues-depth/population/index.html (accessed December 14, 2020).

White, H. (2011) 'Power shift: Rethinking Australia's place in the Asian century', *Australian Journal of International Affairs*, 65 (1): 81–93.

2

WHAT'S NEW IN URBAN DATA ANALYTICS?

Elisabete A. Silva, Lun Liu, Heeseo Rain Kwon, Haifeng Niu, Yiqiao Chen and Jose Reis

Introduction

Modern societies necessitate high levels of complexity in order to properly function; systems are interrelated, creating systems of systems, and the sum of all propels the society forward. Therefore, cities and human societies are regarded as complex entities (Cao and Silva, 2017; de Roo and Silva, 2010). Uncertain, unpredictable and complex urban problems require the urban planning systems to adapt and evolve in order to cope with fast changes such as climate changes, traffic accidents and health crises. As a consequence, the theoretical paradigms associated with urban planning needed to evolve, too. Nowadays complexity theory and adaptive planning supported by new data science approaches seem to be at the centre of cities' research and practice. Instead of regarding cities as still-static planning systems, complexity theory takes cities as comprehensive entities, as dynamic intertwining systems, and as systems of systems that are self-organising and permanently changing. These cities require alternative forms of decision-making that can accommodate their ever-changing nature as almost live organisms (de Roo and Silva, 2010).

In response to the high levels of integration required, adaptive planning is proposed as a more dynamic approach together with complexity theory. The objective of adaptive planning is to manage various potential development trajectories while promoting preferred outcomes, based on the understanding of the uncertainties in urban development (Rauws and de Roo, 2016; Silva, 2010). In these scenarios, the concept of smart city has been applied in planning in order to handle the growing complexity and to build a better adaptive planning system (de Roo and Silva, 2010). In smart cities, information and communication technology (ICT) is incorporated in infrastructures, architectures and everyday objects, and enables the generation of vast data sets for planners. Such data sets can provide planners with a clearer picture and up-to-date reality so that a more adaptive planning system can be built to address comprehensive urban problems (Townsend, 2014).

Urban data analytics is an essential part of the concept of smart city and adaptive planning. It refers to 'techniques that are used to analyse and acquire profound knowledge out of urban data' (Moustaka et al., 2019: 7). Conventional data in urban analysis contains census data, housing survey, land cover official data and post office data which are sometimes expensive, uncertain, inaccurate, and time-consuming (Brito-Henriques et al., 2018). Nevertheless, regular frequency of data collection and methodological consistency are still one of the best attributes of such traditional data. Conventional data can serve as the foundation analytics for not only the analysis of more refined data for validation and calibration, but also for time analysis and examination of long range trends (Silva et al., 2020). However, tasks that require instant response and management such as real-time traffic management, recognition of emergency response activities (e.g. earthquake, flood), and pollution monitoring are not likely to be solved by conventional data only (Chen and Silva, forthcoming). Nowadays, numerous sensors, online posts, digital pictures and other records in smart cities contribute to the explosion of digital data, which is known as 'big data' (Bagiwa, 2017). Big data and data analytics can complement the conventional data to address the medium- and long-term urban issues such as public participation, social segregation, regional connection, and functional division (Chen et al., 2020; Hao et al., 2015).

Despite the great potential of big data and urban data analytics, they have also been criticised for being techno-centric and entrepreneurial, using black box methods, and widening inequalities (Kitchin, 2015; Mora et al., 2017). The criticisms remind us of the seven sins presented by Lee (1973) forty years ago, namely 'hypercomprehensiveness', 'grossness', 'wrongheadedness', 'hungriness', 'complicatedness', 'mechanicalness', and 'expensiveness'. While new data sources, analytical tools and modelling approaches have been partially effective in addressing these criticisms (Batty, 2014; Te Brömmelstroet et al., 2014), challenges that we identify in this chapter indicate that some of these sins suggested by Lee (1973) remain as valid criticisms for urban computation and urban data analytics.

Against this background, in the data rich environment, extremely large data sets raise the questions of how to collect, manage, and integrate diverse types of data in diverse formats and scales (Silva et al., 2020). Currently, planning systems worldwide are mostly static, where the duration of data collection and analysis takes years, making it difficult to achieve the goals in a new smart city environment in order to manage the complexities at hand.

With the methods of new data analytics and vast computation power, it is possible to provide potential solutions for the challenges of collecting big data for adaptive planning. With the help of ICT advances and data science, we are now able to collect spatiotemporal data from the data rich environments we now live in and produce smarter cities that adapt sooner and quicker to the challenges at hand (Bibri, 2019; Kitchin, 2014). Crowdsourced data collection, as a bottom-up data collection process, is one of the examples. Crowdsourced data consist of social network services, point of interest and collaborative websites, which are mostly georeferenced or geotagged. Most of online crowdsourced data provides an 'application programming interface' (API) which allows application developers to collect data from the website database, usually needing only basic computer programming skills (Silva et al., 2020). The new data collection techniques are widely available at low cost.

This chapter examines the latest issues of urban data analytics in the following five challenges: the data collection challenge, the data mining challenge, the data integration challenge, the time and space challenge, and the standards/norms challenge. It finishes with a discussion named 'the policy support systems challenge', which integrates all five challenges, analyses how some of the concerns raised by Lee (1973) are still applicable, and points to future directions.

The data collection challenge

We live in a data rich environment, where many data sets are harvested from various forms of sensors around the city. Many cities have deployed thousands of sensors that monitor the air, sounds, heat, building structures, as well as vehicle and human flows in the city. It thus poses a big challenge to collect, manage and integrate such diverse types of data (formats, scales, noise, etc.), and a more intellectual challenge to generate real information and insights from the large volume of data.

The family of urban sensors has expanded considerably in recent years from traditional remote sensors and hyperspectral sensors, to those prompted by new technologies and services such as the in-car Global Positioning System (GPS) that send real-time locations of vehicles and Street View Cars that routinely collect street view imagery (Liu et al., 2017). There are multiple ways of categorising existing urban sensors, for example, dedicated sensors and non-dedicated sensors (Habibzadeh et al., 2017).

Dedicated sensors refer to sensors installed for specific urban purposes such as air quality sensors (Dutta et al., 2009) and home energy monitors (Buchanan et al., 2015). On the other hand, non-dedicated sensors refer to instruments or activities implemented primarily for other purposes but then are alternatively used as urban sensors. The examples include public transit cards which were originally for payment, but are now also used to monitor the flow of people (Lathia and Capra, 2011) and cameras for capturing car number plates in the congestion charging zone that were originally for tolling, but are now also used to monitor the vehicle flow (Lawlor et al., 2016).

Alternatively, sensors can also be categorised as direct instrumenting, which involves the actual deployment of instruments, and indirect sourcing such as the crowdsourcing of urban information based on individuals' contributions and various traces left by human activities on the internet (e.g. social media posts in Dodds et al., 2011). In more systematic programs of urban sensing, sensors are further linked to control centres to feed real-time information, which forms the digital infrastructure for the smart city vision.

While it is true that sensors provide unprecedented information about what is going on in the city, it still requires a lot of intelligence from researchers and practitioners to mine the data to inform policy-making. A primary issue is how to link separate data sets and draw a whole picture of the running of cities, since any single data set, while very much detailed, usually shows only the dynamics in one sector of the urban system instead of correlations and causations. For instance, for road congestion, we might have the trajectory data of a number of vehicles around (via in-car GPS, etc.) or the dynamics of traffic flows at road crossings (via vehicle detector loops, etc.), yet we might also need to investigate the functions and activities on surrounding lands to achieve a deeper understanding of causes and thus design prevention measures. Besides, to make right use of sensor data, the unevenness and bias in the data also need to be carefully dealt with. It is tempting to take what the sensors reveal as the truth, yet it is not always the case. For instance, the crowdsourced data of OpenStreetMap is found to be of high quality in London, but not in many other cities (Haklay, 2010) and the records of 311, the most widely used municipal service in the U.S., are found to be less representative for minority groups (Townsend, 2013).

Moreover, before stepping into data analytics, one should also be aware that not all the starring data are freely accessible and not all data accessible are legible (and legal) to use and publish, despite of the 'open data era' we are now in. There are data sets that are both big and open, and data that are big but private (Gurin, 2014), which may be held by governments, enterprises or research institutes for the sake of property right, commercial regulations or privacy concerns. More discussion of the related legal and ethical issues is provided later in this chapter when addressing 'the norms challenge'.

The data mining challenge

In the data rich environment, the average size of an urban data set can be in the terabyte range. Data mining, as an interdisciplinary concept from computer science and statistics, has become a key process in information discovery from big data sets (action supported by powerful algorithms based on mathematical and statistical models and machine learning). In the field of urban analytics, the task of data mining mainly focuses on discovering urban patterns and regularities from spatial or spatiotemporal data sets (Miller and Han, 2009). However, with the development of ICTs and sensors, urban data sets now have more features beyond space and time, which brings challenges to data mining methods in terms of dealing spatial-temporal features with others.

Both statistical/mathematical models and machine learning are applied as data mining methods for spatial patterns (Cheng et al., 2014). Statistical models such as spatial autoregression model (SAR) and geographically weighted regression (GWR) provide tools to explore spatial relationships and model the patterns of spatial distribution. However, for statistical models, it is challenging to select explanatory variables from new features in big urban data sets and to capture non-linear relationship between variables. In comparison, machine learning algorithms for *clustering, classification* and *regression* provide good support for pattern identification and predictions from large data sets (Han et al., 2011). Although machine learning performs well in mining new urban data sets, machine learning algorithms fail to provide usable interpretations of the patterns identified, which is a huge challenge (Miller and Goodchild, 2015).

In addition to analytical challenges, another challenge in data mining is related to information extraction with algorithms. Structured data, such as administrative data, geographic-referenced maps, and environmental data stored either in tabular format or relational databases, are commonly processed in urban analytics. However, there are increasing amounts of unstructured data such as text, audio, and images generated from social media platforms and information hubs everyday via all kinds of sensors. This type of data not only have fine-scale space and time tags but also information about humans and their surrounding spaces in cities. We have started to see some attempts of unstructured data mining for measurement of urban environment, sentiment analysis and so on (Niu and Silva, 2020). For instance, Liu et al. (2017) measured a large-scale street environment in Beijing using street view images and computer vision algorithms, and explored its relationship with urban activities. Chen et al. (2020) measured the sentiment feedback about regrowth-planning policy with natural language processing tools. With more and more unstructured data generated in our cities, we believe that leveraging unstructured data with new data pre-processing methods will be a hot topic and a big challenge in future work.

The data integration challenge

In a world of vast amounts of sensors, terabytes of information, and second-by-second data flow, it is another big challenge to integrate data that are collected in different platforms and represented in different forms. For instance, many urban-related data sets are spatial and are generally divided into raster (e.g. from satellite or remote sensing) and vector data structures (e.g. from web-based map services). There is also a challenge in inter-operating temporal and spatial dynamic structures, which involves time evolution, scalability, immersive realities and beyond. These conditions make the data at hand vast but sometimes messy (Harford, 2014; Kitchin, 2014).

A straightforward classification of data structures is dividing them into spatial and aspatial data. Spatial data are generally dominant in the field of urban data analytics since most urban events have locations associated to them. However, the resolution of spatial information can vary substantially: from coordinate-level (e.g. GPS data) to coarser resolutions of postcode-level (e.g. some governmental data) to street-level (e.g. cell phone signal, vehicle detector camera or coils at street crossings), district-level, to city-level (e.g. census data), and to regional and even country level. The frequency of update is also an important aspect of the data structure. In this sense, urban data can be classified along the spectrum of update frequency: from real-time (e.g. most sensor data and the records of online activities), to weekly, monthly and annually updated data.

Another issue that is drawing growing concern is the treatment and mining of unstructured data such as texts, images, audios and videos, which lack the structural organisation required by machines for analysis in contrast to structured data that are in tidy formats of tables or vectors (Gandomi and Haider, 2015). The recent progress in the state-of-the-art machine learning algorithms is now opening new opportunities for the distilling of unstructured data. For instance, by applying machine/deep learning models to street view images, we can now measure several features of the urban environment that were hard to measure in the past (Liu et al., 2017).

The large diversity in data structure raises many operational questions in urban data analytics. It is very likely that the multiple data sets used in one analysis do not align with each other, either temporally or spatially. The data might be produced at different time points, using different temporal intervals, at different spatial resolution and with different spatial boundaries. This is to an extent similar to the classic Modifiable Areal Unit Problem which refers to the inconsistency of analytical results due to different spatial configurations or partitioning schemes (Openshaw, 1983; Viegas et al., 2009; Wong, 2009). However, it is far less clear how analysis results would be affected by different approaches of manipulating and aligning the highly diverse data. Besides, in many circumstances, highly disaggregated data need to be clustered to align or synchronise with each other, yet such a treatment might also result in the loss of some detail as well as the distortion of data (Briant et al., 2010).

To fill in this gap, we should expect more research in the future that evaluates this problem of inconsistency and the consequences of different approaches of data treatment. Hopefully, there will be more protocols about how data can be combined and when it is likely to induce errors, for example, how large the temporal gap between two data sets can be for them to be used in one analysis. This will depend on the content of the data sets. Moreover, considering that combining different data sets usually involves scaling up and losing details, research into innovative ways of data treatment would also be helpful.

The time and space challenge

The fourth challenge in urban data analytics is the time and space challenge – the need for dynamic spatial metrics and models to capture the interaction of time and space. Spatial metrics are quantitative measures that evaluate the physical characteristics of geographical features and can be used to understand the spatiotemporal patterns of urban development (Reis et al., 2016). Coupled with dynamic spatial urban models, such as cellular automata (CA) and agent-based models (ABM), spatial metrics with temporal variability can be used to simulate different planning scenarios to explore the expected impacts of different policy decisions. Such dynamic simulations can then play an important role in planning policy support by aiding the decision-making process, community participation and stakeholder engagement, especially with the increase of people's ICT literacy (Falconi and Palmer, 2017; Le Pira et al., 2017). However, existing studies tend to address space and time separately when modelling temporal dynamics – processing static metrics at multiple time intervals rather than achieving a true integration of time and space together with aspatial data. We work with dynamic models using very detailed data and short timeframes, but the metrics to do all this work are still the same we used fifty years ago. There are very few dynamic metrics in our dynamic models representing this very fast moving dynamic world. Dynamic metrics consider 'time' along with the usual spatial dimension, for example, 'physical-demographic mismatch metric' developed by Reis et al. (forthcoming) which represents the temporal change in the ratio between residential and population density. Such integration is challenging in a world of quasi-live big data with thousands of local and remote sensors and with terabytes of information collected on a regular basis and becoming obsolete at the same speed.

This time and space challenge has been tackled by a number of researchers with various state-of-the-art methods of spatiotemporal data exploration. One is the concept of three-dimensional 'blob' introduced by Parrott et al. (2008) where successive time-series spatial raster images are stacked in a space-time cube, enabling metrics such as number of blobs, density, blob shape complexity, spatiotemporal complexity, fractal dimension, and contagion (Bach et al., 2017; Bhatti et al., 2018). Another is the hybrid approach of ABM and CA to allow simulation of individual behaviour in three-dimensional model. Koch et al. (2019) created a hybrid ABM-CA approach to model dynamic landowner interactions and development patterns at the urban fringe and concluded that peer-influence intensities indirectly affect the patterns of urbanisation and fragmentation of undeveloped land. Wu and Silva (2014) produced a similar ABM-CA integration approach with the use of transition tables between data structures. Furthermore, developments are being made in time geography studies to deal with the spatiotemporal big data emerging in urban environments. Chen et al. (2016) proposed a spatiotemporal data model based on a compressed linear reference (CLR) technique, which enables the storing and management of network time geographic entities in classic spatial databases. They presented a prototype and a case study using GIS techniques and large-scale data sets of space-time paths and prisms (Chen et al., 2016). Another example is the research by Xiao et al. (2020) which integrated Hägerstrand's time geography with data mining and big data analysis. Xiao et al. (2020) examined mobile phone signal data to generate individuals' origin-destination flows in Shanghai and drew out important findings about the job-housing imbalance that overtime workers face as well as the interchangeability of the temporal and spatial exploitations of labour.

Other approaches of spatiotemporal analysis include hidden Markov models (HMM) and computational fluid dynamics (CFD). HMM can simulate time-series observations and has been successful in examining urban growth and land use dynamics (Gong et al., 2017; Singh et al., 2015); however, they tend to rely on the implementation of dynamic models rather than dynamic metrics. CFD is another three-dimensional model that can be applied to handle temporal dynamics with the spatial domain (Blocken, 2015; Chang et al., 2018). However, both models consider each time step as a separate layer and do not integrate dynamic metrics to achieve full optimisation (Bhatti et al., 2018).

The standards/norms challenge and the regulatory/legal imperative

Standardisation of data collection is pivotal for reducing uncertainty and errors in data analysis, as well as for enabling data integration and scalability of analysis. Institutions such as the British Standards Institution (BSI) and the International Organization for Standardization (ISO) play the role as key beacons of good practice around the world. According to the BSI (2019), a 'standard' is 'an agreed way of doing something'. Examples of standards in urban data analytics include ISO 19115-1: 2014 (Geographic information - Metadata - Part 1: Fundamentals), ISO 6709: 2008 (Standard representation of geographic point location by coordinates), ISO 19101-1: 2014 (Geographic information - Reference model - Part 1: Fundamentals), ISO 19111: 2019 (Geographic information - Referencing by coordinates), and ISO/TS 19130-2: 2014 (Geographic information - Imagery sensor models for geopositioning - Part 2: SAR, InSAR, LiDAR and sonar). These are only a few of thousands of ISO norms that, for instance, allow us to rely completely on devices that take our car from point A to point B without getting lost and allow infrastructure workers to repair water pipes in the pavement without interfering with other infrastructure.

The 'real world' regulations, laws, and norms can define the limits of what we can and cannot do, for example, road and building regulations that keep us safe. In this context, and considering data science, big data, geographic information and ICT, it is fair to expect that the 'digital world' can be equally regulated. Government regulations, such as the Data Protection Act 2018 of the United Kingdom, are extremely important for expanding security and safety measures from the real world to the digital world (Data Protection Act 2018 Explanatory Notes).[1] Regulation sets standards for the processing of personal data about data subjects (i.e. identified or identifiable natural person) by data controllers (i.e. organisations). Sensitive personal data such as racial or ethnic origin, political opinions, religious beliefs and health condition get even greater protection. The awareness of digital privacy also brings changes in data collection and processing in urban analytics. Although the introduction of regulations has increased the public awareness of digital privacy, the ethical dilemma still exists.

With the rise of public awareness of data privacy, the dilemma of using new urban big data for urban analytics becomes more serious. It is still not entirely clear how to properly use highly disaggregated and personal data such as micro-economic data and health records to meet the public and private interests in strategic and operational planning. For instance, the government needs deep learning technologies from the high-tech companies to analyse citizens' health records for better planning of limited healthcare resources. How to better provide public goods in cooperation with the private sector while protecting citizens' privacy is still and will continue to be a problem. Another dilemma is that current regulation cannot effectively protect personal information, which only emphasises data anonymisation. This occurs because, by cross-referencing other location information with time tags, anonymous data of individuals' activities can be re-identified. Even sensitive personal information such as socio-demographic information can be inferred or matched just by a few other low-priority data points (Rocher et al., 2019). Those dilemmas prove that there is a huge gap in the regulation of new big data to balance privacy and public interests.

To close the gap, firstly, we need to revisit the roles of the public and private sectors in data collection, processing, management and the collaboration between them. The legislation needs to create different levels of access to information depending on the need and on the security level that it requires. On top of this, western cities need to start promoting the government as the 'guardian of data'. Criticisms tend to be directed at the public sector often in a disproportionate manner considering that data collection and privacy issues exist at a large scale, perhaps larger scale, in the

---

[1]Data Protection Act 2018 Explanatory Notes Queen's Printer of Acts of Parliament. http://www.legislation.gov.uk/ukpga/2018/12/notes (accessed December 15, 2020).

private sector (Ryan and Gregory, 2019). The public sector, especially in those cities where public functions have been outsourced, needs to directly collect data sets that are associated with public interests such as tax and health-related data (Information Commissioner's Office, 2019). Breaking up the monopolies of new urban data collected/provided by the private sector can in turn provide support for policy-making and planning, and effectively provide public goods during the public emergency and crisis. Secondly, the future regulations should not only set standards about how citizen data are stored in the input stage, but also regulate how their data are processed and what are the purposes of the analytics to avoid sensitive information exposition by cross-referencing other data or inferencing other related data with machine learning.

Conclusion

In a time of big data, new data sources and urban analytics have become hot topics in urban planning and triggered a new trend of quantitative research. However, we cannot be overly optimistic about the potential of big data, data analytics, and urban simulation as we have seen the decline and failure of urban computation in history. The year 1973 was an important year for those involved in urban computation to remember as it marks the publication of Lee's 'Requiem for Large-Scale Models' (1973). This publication put a stop to most of the effort of urban modelling in subject areas such as geography, planning, and architecture and marked a shift towards a more qualitative approach in university departments around the world. This was also associated with a paradigm shift of urban policy towards more qualitative research (Batty, 2014; Te Brömmelstroet et al., 2014).

With the rise of big data and the establishment of smart cities, it is time to stop and re-evaluate, not only the opportunities available but also the perceived threats. With that goal, it makes sense to start the basis of such self-assessment by using Lee's 'Requiem for Large-Scale Models' that collapsed the comprehensive modelling efforts in urban planning forty years ago (Lee, 1973). Lee stated seven sins in large-scale models, in which 'grossness', 'hungriness', and 'expensiveness' were in the data realm. 'Grossness' meant that excessive data in the models provided micro perspectives while policymakers need macro views. The sin of 'hungriness' referred to the great data requirement in building large-scale data. 'Expensiveness' can be seen as the cost of building models and the cost of collecting various data sets (Lee, 1973).

When Lee wrote about the sin of 'hungriness' for data forty years ago, he might not have imagined the richness of data that we are enjoying today. The easy availability of data, which are now supplied via remote and local sensors instead of painstaking local collection, makes the usage of a large amount of data less of a sin compared to Lee's time. The availability of big data and the advancement in data analytics in recent years have solved 'grossness', 'hungriness', and 'expensiveness' to some extent; yet, these 'sins' are still perceived as threats and the arguments are still very valid.

Today's data visualisation techniques combined with detailed information in simulation, especially using CA and ABM approaches, can provide better interfaces for policymakers and the public to understand the models. However, 'grossness' has not been addressed because many refined models are still less operational for policy purposes (Batty, 2014). The data rich environments nowadays allow urban modellers to have access to vast amount of the data that they need, which can address 'hungriness'. The open access movement and more advanced data collection techniques such as crowdsourced collection have reduced the cost of data to a large extent. Nevertheless, some commercial data sets are expensive and the large data pool has increased our hunger for more data (Te Brömmelstroet et al., 2014). While more governmental data are open and the crowdsourced big data can be inputted into research, the sin of 'expensiveness' remains because not all data can be accessed freely. While models at Lee's time were developed at costs of around 500,000 USD, smart city projects nowadays are granted tens of millions of dollars (e.g. 24 million GBP for Glasgow's trial in 2013[2] and 11.1 million USD for Santander in 2014).[3] In many cases, especially in the private sector, the big data sets themselves are considered valuable products, traded as key commodities.

In the current evolution of urban modelling, the process of data mining becomes more and more important. While the background of the 'Requiem for Large-Scale Models' stated by Lee (1973) has completely changed with the development of computer science and the explosion of big data, some challenges remain. Apart from the challenges of high-dimensional attributes and heterogeneity of data sets discussed above, another challenge related to machine learning model choice needs attention. Although the efficiency and prediction performance of black-box models are better than white-box models, urban modelling should still use them with caution because black-box models should not have an impact on policy without proper interpretation. The future of modelling will be based on machine learning techniques, and that trajectory can already be seen as they are used by urban practitioners in the transport, real estate prediction, tourism, and planning industries (Chaillou et al., 2017; Woods et al., 2017). For example, market trends of real estate can be better and faster predicted using deep learning algorithms; however, the reputation of the results is an important issue because investors may not trust the black box (Chaillou et al., 2017). This issue of reliability of machine learning results can only be fully resolved when the results are properly explained. This requires knowledgeable researchers with very good expertise in data mining and model development who are simultaneously aware of the importance of policy and the contributions of planning support systems as a supporting tool to decision makers. For example, Chen et al. (2020) raised the question of whether the public is satisfied with the new planning policy. They answered the question with sentiment analysis, supported by natural language processing (an approach in machine learning), and proper interpretation of the results, with the understandings of the city, the planning process, and public opinion (Chen et al., 2020).

---

[2]https://www.bbc.co.uk/news/technology-21180007 (accessed December 15, 2020).

[3]http://www.governing.com/topics/urban/gov-santander-spain-smart-city.html (accessed December 15, 2020).

Furthermore, the use of dynamic metrics and models enabled by the new data rich world can act as an important breakthrough in addressing the seven sins of large-scale models suggested by Lee (1973). What seemed to be 'hypercomprehensiveness', 'complicatedness', and 'mechanicalness' in the past is not necessarily the case anymore. Integrating time and space with aspatial data using large data sets and high computing power can create a model that sufficiently replicates a complex urban system and serves multiple purposes. Also, the criticism on the coarseness of data ('grossness'), 'hungriness' and 'expensiveness' can be overcome by the widely available big data today with almost microscopic detail even in the temporal scope (e.g. minute-by-minute movement of all individual vehicle passing through sections of the road), as well as the widely available human resources for computer modelling compared to the past. Furthermore, such detailed data make it possible to overcome the criticism of 'wrongheadedness'. It has become not only possible to impute individual behaviour directly from highly disaggregate data, but also to make predictions for the future based on the dynamic patterns of change in the past few years (Kwon and Silva, 2020; forthcoming).

The availability of large data sets and high computing power enables an inclusion of 'complexity' in urban models that was previously not possible. Subsequently, complexity theory should be taken as the overall theoretical concept for dynamic simulation (de Roo and Silva, 2010). However, Lee's Requiem (1973) still offers important insights that modellers of dynamic spatiotemporal analysis should keep in mind, perhaps even more so in the era of large data availability. Narrowing down the scope and setting key questions that the model should answer are ever more important because modellers can get easily lost in vast amount of data. Also, despite the dramatic advancement of computing power, big data and variables will always need to be used selectively to only include spatial, aspatial, and temporal information which is sufficiently relevant to the research questions to prevent unnecessary complication. The development of both static and dynamic metrics should continue to be used as the framework to base spatial models on.

The challenges and concerns of Lee's seven sins of urban modelling should also be extended to the 'sins' of the current planning policy and processes in place. The 'static' characteristic of the planning system in the UK (and many other countries) has not changed much in the past 50 years and it is largely based on immovable masterplans set for a future of around 25 years with static snapshots in time, relying on data collections that happen every 5 years. Not only this, it takes a long while, often around 5 years, for the current planning system to move from data analysis to discussion and to policy implementation. How can such a static system cope with and appropriately utilise fast-moving data collection provided by thousands of sensors, terabytes of information, and multiple data structures in a new smart city environment that requires up-to-date information in order to manage this complex world?

Urban data analytics is closely linked with this paradigm shift from static to adaptive planning. This involves making master plans capable of adjusting to fast-changing realities that cannot wait for 5 years to receive policy approvals. The approach of rational planning and absolute certainty in planning for the next 25 years should be replaced by an incremental process that can reflect changes, which can even be in real-time in some areas if appropriate. This requires policies that grant certain flexibility achieved through 'ranges' of suitability instead of an approach of 'one number/policy fits all'. How can we achieve short timeframes in the spatial planning system in the future similar to how the timelines are currently adjusted appropriately in certain sectors like emergency, disaster, security, and safety planning? How can cities utilise big data to immediately reflect on policies? Examples in the transport sector can be adjusting bus routes, frequency, and service time based on big data analysis and, furthermore, offering appropriate services to the right locations of hotspots of people using data such as location-based mobile signal data (e.g. Xiao et al., 2020).

How can we make sure that data availability, access and policy development are not compromised in an increasingly privatised world? The outsourcing of services to the private sector and the necessary public-private partnerships seemed to forget data in the terms of their contracts. Governments and citizens are more and more realising that, while open data efforts are positive, it is difficult to request access to the data produced by private firms involved in the delivery of public services. This realisation is causing some experts to state that in five years, it is going to be difficult for governments to produce data-led public policy as governments will not have the required data to produce the required studies, even if those same governments are paying for the management of those services via outsourcing. We need to start to make sense of the legal, economic and political imperative in urban data.

References

Bach, B., Dragicevic, P., Archambault, D., Hurter, C. and Carpendale, S. (2017) 'A descriptive framework for temporal data visualizations based on generalized space-time cubes', *Computer Graphics Forum*, 36 (6): 36–61. https://doi.org/10.1111/cgf.12804

Bagiwa, L.I. (2017) 'Big data: Concepts, approaches and challenges', *International Journal of Computer Network and Communications Security*, 5 (8): 181–7.

Batty, M. (2014) 'Can it happen again? Planning support, Lee's Requiem and the rise of the smart cities movement', *Environment and Planning B: Planning and Design*, 41 (3): 388–91.

Bhatti, S.S., Reis, J.P. and Silva, E.A. (2018) 'Spatial metrics: The static and dynamic perspective', in B. Huang (ed.), *Comprehensive Geographic Information Systems* (Vol. 1). Oxford: Elsevier. pp. 181–96.

Bibri, S.E. (2019) 'On the sustainability of smart and smarter cities in the era of big data: An interdisciplinary and transdisciplinary literature review', *Journal of Big Data*, 6 (1): article #25.

Blocken, B. (2015) 'Computational fluid dynamics for urban physics: Importance, scales, possibilities, limitations and ten tips and tricks towards accurate and reliable simulations', *Building and Environment*, 91: 219–45. doi.org/10.1016/j.buildenv.2015.02.015

Briant, A., Combes, P.P. and Lafourcade, M. (2010) 'Dots to boxes: Do the size and shape of spatial units jeopardize economic geography estimations?', *Journal of Urban Economics*, 67 (3): 287–302.

British Standards Institution (2019) What is a standard? https://www.bsigroup.com/en-GB/standards/Information-about-standards/what-is-a-standard/?awc=3911_1614081533_c329ed26e9e3337f3c29327114011a3b (accessed December 10, 2019).

Brito-Henriques, E., Morgado, P. and Cruz, D. (2018) 'Morfologia da cidade perfurada: Padrões espaciais de ruínas e terrenos vacantes em cidades Portuguesas [Morphology of the perforated city: Spatial patterns of ruins and vacant land in Portuguese cities]', *Finisterra-Revista Portuguesa de Geografia*, 108: 111–33.

Buchanan, K., Russo, R. and Anderson, B. (2015) 'The question of energy reduction: The problem(s) with feedback', *Energy Policy*, 77: 89–96.

Cao, K. and Silva, E.A. (2017) *Comprehensive Geographic Information Systems: Vol. 3. GIS Applications for Socio-Economics and Humanity*. Oxford: Elsevier.

Chaillou, S., Fink, D. and Gonçalves, P. (2017) 'Urban tech on the rise: Machine learning disrupts the real estate industry. Featuring interviews of: Marc Rutzen and Jasjeet Thind by Stanislas Chaillou, Daniel Fink and Pamella Gonçalves', *Field Actions Science Reports: The Journal of Field Actions*, (Special Issue 17): 38–43.

Chang, S., Jiang, Q. and Zhao, Y. (2018) 'Integrating CFD and GIS into the development of urban ventilation corridors: A case study in Changchun City, China', *Sustainability*, 10 (6). doi.org/10.3390/su10061814

Chen, B.Y., Yuan, H., Li, Q., Shaw, S.-L., Lam, W.H.K. and Chen, X. (2016) 'Spatiotemporal data model for network time geographic analysis in the era of big data', *International Journal of Geographical Information Science*, 30 (6): 1041–71. doi.org/10.1080/13658816.2015.1104317

Chen, Y. and Silva, E.A. (forthcoming) 'How can complexity theory and data science assist smart city governance? A review', manuscript under revision.

Chen, Y., Silva, E.A. and Reis, J. (2020) 'Measuring policy debate in a regrowing city by sentiment analysis using online media data: A case study of Leipzig 2030', *Regional Science Policy & Practice*. doi.org/10.1111/rsp3.12292

Cheng, T., Haworth, J., Anbaroglu, B., Tanaksaranond, G. and Wang, J. (2014) 'Spatiotemporal data mining', in M.M. Fischer and P. Nijkamp (eds), *Handbook of Regional Science*. Berlin: Springer. pp. 1173–93.

de Roo, G. and Silva, E.A. (2010) *A Planner's Encounter with Complexity*. Farnham: Ashgate.

Dodds, P.S., Harris, K.D., Kloumann, I.M., Bliss, C.A. and Danforth, C.M. (2011) 'Temporal patterns of happiness and information in a global social network: Hedonometrics and Twitter', *PloS One*, 6 (12): e26752.

Dutta, P., Aoki, P.M., Kumar, N., Mainwaring, A., Myers, C., Willett, W. and Woodruff, A. (2009) 'Common sense: Participatory urban sensing using a network of handheld air quality monitors'. SenSys '09: Proceedings of the 7th ACM conference on embedded networked sensor systems. pp. 349–50. doi.org/10.1145/1644038.1644095

Falconi, S.M. and Palmer, R.N. (2017) 'An interdisciplinary framework for participatory modeling design and evaluation – What makes models effective participatory tools?', *Water Resources Research*, 53 (2): 1624–45. doi.org/10.1002/2016WR019373

Gandomi, A. and Haider, M. (2015) 'Beyond the hype: Big data concepts, methods, and analytics', *International Journal of Information Management*, 35 (2): 137–44.

Gong, W., Fang, S., Yang, G. and Ge, M. (2017) 'Using a hidden Markov model for improving the spatial-temporal consistency of time series land cover classification', *ISPRS International Journal of Geo-Information*, 6 (10): article #292. doi.org/10.3390/ijgi6100292

Gurin, J. (2014) 'Big data and open data: How open will the future be?', *I/S: A Journal of Law and Policy* 10 (3): 691–704.

Habibzadeh, H., Qin, Z., Soyata, T. and Kantarci, B. (2017) 'Large-scale distributed dedicated- and non-dedicated smart city sensing systems', *IEEE Sensors Journal*, 17 (23): 7649–58.

Hägerstrand, T. (1970) 'What about people in regional science?', *Papers in Regional Science*, 24: 7–24.

Haklay, M. (2010) 'How good is volunteered geographical information? A comparative study of OpenStreetMap and Ordnance Survey datasets', *Environment and Planning B: Planning and Design*, 37 (4): 682–703.

Han, J., Pei, J. and Kamber, M. (2011) *Data Mining: Concepts and Techniques*. Boston: Morgan Kaufmann. pp. 1–38.

Hao, J., Zhu, J. and Zhong, R. (2015) 'The rise of big data on urban studies and planning practices in China: Review and open research issues', *Journal of Urban Management*, 4 (2): 92–124.

Harford, T. (2014) 'Big data: A big mistake?', *Significance*, 11 (5): 14–19.

Information Commissioner's Office (2019) *Outsourcing Oversight? The Case for Reforming Access to Information Law*. London: Information Commissioner's Office. https://ico.org.uk/media/about-the-ico/documents/2614204/outsourcing-oversight-ico-report-to-parliament.pdf (accessed December 15, 2020).

Kitchin, R. (2014) 'The real-time city? Big data and smart urbanism', *Geojournal*, 79 (1): 1–14.

Kitchin, R. (2015) 'Making sense of smart cities: Addressing present shortcomings', *Cambridge Journal of Regions, Economy and Society*, 8 (1): 131–6.

Koch, J., Dorning, M.A., Van Berkel, D.B., Beck, S.M., Sanchez, G.M., Shashidharan, A., Smart, L.S., Zhang, Q., Smith, J.W. and Meentemeyer, R.K. (2019) 'Modeling landowner interactions and development patterns at the urban fringe', *Landscape and Urban Planning*, 182: 101–13.

Kwon, H.R. and Silva, E.A. (2020) 'Mapping the landscape of behavioral theories: Systematic literature review', *Journal of Planning Literature*, 35 (2): 161–79. doi.org/10.1177/0885412219881135

Kwon, H.R. and Silva, E.A. (forthcoming) 'Matching behavioral theories and rules with spatial planning methods', manuscript under revision.

Lathia, N. and Capra, L. (2011) 'How smart is your smartcard? Measuring travel behaviours, perceptions, and incentives'. Paper presented at the Proceedings of the 13th international conference on Ubiquitous computing, Beijing, China, September 17–21. https://www.researchgate.net/publication/221568348_How_smart_is_your_smartcard_Measuring_travel_behaviours_perceptions_and_incentives (accessed December 15, 2020).

Lawlor, S., Sider, T., Eluru, N., Hatzopoulou, M. and Rabbat, M.G. (2016) 'Detecting convoys using license plate recognition data', *IEEE Transactions on Signal and Information Processing over Networks*, 2 (3): 391–405.

Le Pira, M., Marcucci, E., Gatta, V., Ignaccolo, M., Inturri, G. and Pluchino, A. (2017) 'Towards a decision-support procedure to foster stakeholder involvement and acceptability of urban freight transport policies', *European Transport Research Review*, 9 (4): article #54. doi.org/10.1007/s12544-017-0268-2

Lee, D.B. (1973) 'Requiem for large-scale models', *Journal of the American Planning Association*, 39 (3): 163–78.

Liu, L., Silva, E.A., Wu, C. and Wang, H. (2017) 'A machine learning-based method for the large-scale evaluation of the qualities of the urban environment', *Computers,*

Environment and Urban Systems, 65 (Supplement C): 113–25. doi.org/10.1016/j.compenvurbsys.2017.06.003

Miller, H.J. and Goodchild, M.F. (2015) 'Data-driven geography', *GeoJournal*, 80 (4): 449–61. doi.org/10.1007/s10708-014-9602-6

Miller, H.J. and Han, J. (2009) *Geographic Data Mining and Knowledge Discovery* (2nd edition). Boca Raton, FL: CRC Press.

Mora, L., Bolici, R. and Deakin, M. (2017) 'The first two decades of smart-city research: A bibliometric analysis', *Journal of Urban Technology*, 24 (1): 3–27. https://doi.org/10.1080/10630732.2017.1285123

Moustaka, V., Vakali, A. and Anthopoulos, L.G. (2019) 'A systematic review for smart city data analytics', *ACM Computing Surveys*, 51 (5): 1–41. doi.org/10.1145/3239566

Niu, H. and Silva, E.A. (2020) 'Crowdsourced data mining for urban activity: Review of data sources, applications, and methods', *Journal of Urban Planning and Development*, 146 (2): article #04020007. doi.org/10.1061/(ASCE)UP.1943-5444.0000566

Openshaw, S. (1983) *The Modifiable Areal Unit Problem*. Norwich: Geo Books.

Parrott, L., Proulx, R. and Thibert-Plante, X. (2008) 'Three-dimensional metrics for the analysis of spatiotemporal data in ecology', *Ecological Informatics*, 3 (6): 343–53.

Rauws, W. and de Roo, G. (2016) 'Adaptive planning: Generating conditions for urban adaptability. Lessons from Dutch organic development strategies', *Environment and Planning B: Planning and Design*, 43 (6): 1052–74.

Reis, J.P., Silva, E.A. and Pinho, P. (2016) 'Spatial metrics to study urban patterns in growing and shrinking cities', *Urban Geography*, 37 (2): 246–71.

Reis, J.P., Silva, E.A. and Pinho, P. (forthcoming) 'Quantitative assessment of growing and shrinking cities with spatial metrics: Evidence from two Portuguese cities', manuscript under revision.

Rocher, L., Hendrickx, J.M. and de Montjoye, Y.-A. (2019) 'Estimating the success of re-identifications in incomplete datasets using generative models', *Nature Communications*, 10: article #3069. doi.org/10.1038/s41467-019-10933-3

Ryan, M. and Gregory, A. (2019) 'Ethics of using smart city AI and big data: The case of four large European cities', *ORBIT Journal*, 2 (2). doi.org/10.29297/orbit.v2i2.110

Silva, E. (2010) 'Waves of complexity: Theory, models, and practice', in G. de Roo and E. Silva (eds), *A Planner's Encounter with Complexity*. Farnham: Ashgate. pp. 309–31.

Silva, E., Liu, L., Kwon, H.R., Niu, H. and Chen, Y. (2020) 'Hard and soft data integration in geocomputation: Mixed methods for data collection and processing in urban planning', in S. Geertman and J. Stillwell (eds), *Handbook of Planning Support Science*. Cheltenham: Edward Elgar. pp. 37–55.

Singh, S.K., Mustak, S., Srivastava, P.K., Szabó, S. and Islam, T. (2015) 'Predicting spatial and decadal LULC changes through cellular automata Markov Chain Models using earth observation datasets and geo-information', *Environmental Processes*, 2 (1): 61–78.

Te Brömmelstroet, M., Pelzer, P. and Geertman, S. (2014) 'Forty years after Lee's Requiem: Are we beyond the seven sins?', *Environment and Planning B: Urban Analytics and City Science*, 41 (3): 381–7.

Townsend, A.M. (2013) *Smart Cities: Big Data, Civic Hackers, and the Quest for a New Utopia*. New York, NY: WW Norton.

Townsend, A. (2014) 'A manifesto for smart city development', *The European*, March 20. https://www.theeuropean.de/en/anthony-townsend--2/8244-a-manifesto-for-smart-city-development (accessed December 15, 2020).

Viegas, J.M., Martinez, L.M., and Silva, E.A. (2009) 'Effects of the Modifiable Areal Unit Problem on the Delineation of Traffic Analysis Zones', *Environment and Planning B: Planning and Design*, 36 (4): 625–643.

Wong, D.W. (2009) 'Modifiable areal unit problem', in R. Kitchin and N. Thrift (eds), *International Encyclopedia of Human Geography*. Elsevier Inc. pp. 169–74. https://doi.org/10.1016/B978-008044910-4.00475-2

Woods, E., Rodriguez Labastida, R., Citron, R. and Leuschner, P. (2017) 'UK Smart Cities Index 2017', Assessment of Strategy and Execution for the UK's Leading Smart Cities. https://www.milton-keynes.gov.uk/assets/attach/51596/UK-Smart-Cities-Index-2017.pdf (accessed December 15, 2020).

Wu, N. and Silva, E.A. (2014) '2014 DG-ABC: An integrated multi-agent and cellular automata urban growth model', in N.N. Pinto, J.A. Tenedório, A.P. Antunes and J. Roca (eds), *Technologies in Urban and Spatial Planning: Virtual Cities and Territories?* Hershey, PA: IGI Global. pp. 57–92.

Xiao, C., Silva, E.A. and Zhang, C. (2020) 'Nine-nine-six work system and people's movement patterns: Using big data sets to analyse overtime working in Shanghai', *Land Use Policy*, 90: article #104340. doi.org/10.1016/j.landusepol.2019.104340

3

INDICATORS AND URBAN PLANNING: THE BIG DATA OPPORTUNITIES AND CHALLENGES

Cecilia Wong

> The combination of some data and an aching desire for an answer does not ensure that a reasonable answer can be extracted from a given body of data.
>
> (John W. Tukey, 1986: 74–5)

Introduction

The advance of digital technologies and cyber-physical systems has accelerated the importance of artificial intelligence and data science in transforming our analytical capabilities. This chapter aims to examine the opportunities and challenges of 'big data' in relation to urban planning by focusing on the use of quantitative indicators to guide policy-making. Spatial planning is embedded in a dynamic social context with a wide variety of actors and practices, which makes each problem unique but also difficult or even impossible to fully solve. Indicators are surrogates or proxy measures of some abstract, multi-dimensional concepts (Wong and Zheng, 2019); essentially, they are statistics that provide some sort of measurement of a particular phenomenon of concern. A good indicator should be underpinned by rigorous scientific knowledge and able to communicate complex issues in simple terms. Rae and Wong (2012), therefore, advocate a more analytical, adaptive, and spatial approach of using indicators to monitor spatial planning policy. With the rapid pace of big data development, the socio-technical contexts of planning and its monitoring approaches have changed.

With a large amount of data from a whole variety of sources, the expectation in the past was to collect and assemble data to form a consolidated indicator database, on a common spatial and temporal basis, under a clearly identified set of definitions to allow meaningful analysis (Wong, 2006). However, big data is seen as a game-changer because 'computers are for the first time being used en masse to sense change and collect data, while being located remotely and passing their data to other places' (Batty, 2013: 275) to 'allow measurement of city characteristics and outcome variables at higher collection frequencies and more granular geographic scales than ever before' (Glaeser et al., 2018: 114). This requires a fresh look at the changing nature and definition of 'data' against the hype of 'big data'.

Many argue that big data is not new but a slippery concept with a moving definition (Batty, 2013; Press, 2014), as the 'bigness' of data means size matters and is relative to the storage and processing tools at the time (Manyika et al., 2011). The 3 Vs (*volume, velocity* and *variety*) that are commonly used to define big data came from a tweet by Doug Laney (2001, 2012) of the information technology consultant Gartner, Inc. Since then, Gartner has devised a longer definition (Sicular, 2013):[1]

[1]Gartner's most recent definition is 'Big data is high-volume, high-velocity and/or high-variety information assets that demand cost-effective, innovative forms of information processing that enable enhanced insight, decision making, and process automation' by adding process automation to the 2013 definition. (https://www.gartner.com/en/information-technology/glossary/big-data accessed February 10, 2021).

> 'Big data' is high-volume, -velocity and -variety information assets that demand cost-effective, innovative forms of information processing for enhanced insight and decision making.

The increasing *volume* of data is due to the *variety* of low-density, unstructured data from both digital media (e.g. web pages, Twitter, mobile apps and sensor-collected data) and other organisation-held information that is received at a very fast rate (i.e. *velocity*). The emphasis of velocity is also widely interpreted as real-time analytics, but Sicular (2013) provides further clarification of the temporal relationship between data whereby 'velocity is also about the rate of changes, about linking data sets that are coming with different speeds and about bursts of activities, rather than habitual steady tempo'. It is these complexities of big data that present major challenges to data processing and storage capacities and relate to the 'cost-effective, innovative forms of information processing' part of the definition.

The use of quantitative indicators to guide policy-making can be traced back to the 1940s when the monthly economic indicators were first published to measure the buoyancy of the US economy. The quest for indicators has become a worldwide phenomenon since the early 1990s (Innes and Booher, 2000). Entering the key word 'indicators' into Google yielded 653 million entries on February 10, 2021, which is nearly a hundred-fold increase since my last search in 2005 (Wong, 2006). By focusing on urban indicators, 185 million websites appear. It is thus pertinent to question how '*big data*' – with high volume, velocity and variety – can fuel the development of indicators as analytical instruments for urban planning.

After this introduction, the next section provides an overview of the international context of indicator development to inform urban planning, especially against the global context of the UN's *New Urban Agenda* and Sustainable Development Goals (SDGs). It then outlines the long-standing methodological problems of indicator development over the measurement of temporal dynamics and spatial contexts across multiple scales. These challenges are set against a critical discussion over the opportunities and challenges brought by big data to unlock the potential of urban indicator development.

The international urban discourse and development of indicators

Until recently, the international agenda of developing indicators to monitor progress of sustainable development has been very much centred at the national level (Sustainable Cities International, 2012). The United Nations Commission on Sustainable Development (UNCSD, 1996) took a lead by publishing the *Indicators of Sustainable Development: Framework and Methodologies* report in August 1996. The Millennium Development Goals (MDGs) were then agreed by world leaders for poverty eradication, though environmental sustainability was also emphasised. Building on the MDGs, 17 SDGs were agreed to by 193 world leaders in 2015 (United Nations, 2015) and 231 unique indicators[2] were developed to monitor the 169 targets[3] of the SDGs for the 2030 sustainable development agenda (UNDESA, 2017).

With the rapid pace of urbanisation in developing countries (UNDESA, 2012), a consensus has emerged from the international community that it is at the urban level where political leaders, planners and local residents mainly interact and benefit from policy co-production. Cities are seen as clusters of socio-economic activities that directly interact with the natural environment. This was affirmed in the UN's *New Urban Agenda*: cities will be the driver to end poverty, develop sustainable and inclusive urban economies and deliver environmental sustainability (United Nations, 2016). The Global Future Council on Cities and Urbanisation was also established by the World Economic Forum to address both challenges and opportunities brought by urbanisation. The European Commission has changed its orthodox regional development approach by renaming its Directorate-General as 'Regional and Urban Policy' to embrace cities in its agenda. A new type of urbanisation – human-centred and sensitive to environmental outcomes – was advocated in China's 2014 National New Urbanisation Plan (Wong et al., 2018). The Plan introduces 18 key indicators to monitor its urbanisation level, basic public services, infrastructure, and resources and environment to ensure the integration of 'ecological civilisation' into the urbanisation process.

This 'urban turn' has triggered parallel efforts towards developing indicators to monitor progress of goals and targets that fit with the local urban context. One notable effort was the introduction of the City Prosperity Index (CPI) by the United Nations[4] in 2012 to measure productivity, infrastructure, quality of life, equity and inclusion, environmental sustainability, and governance and legislation; each of these is measured by a number of indicators or sub-indicators (Wong, 2015). The CPI set out to move away from 'GDP fetishism' towards measuring the broader conception of human and societal well-being in a holistic and integrative manner, which has also been adopted as a global monitoring framework for the United Nations' SDG11-Sustainable Cities and Communities[5] and its *New Urban Agenda* (United Nations, 2016). Meanwhile, China's National New Urbanisation Plan[6] consists of a set of eighteen indicators measuring urbanisation level, basic public services, infrastructure, and resources and environment (China State Council, 2014).

---

[2]UN Sustainable Development Goal Indicators: https://unstats.un.org/sdgs/indicators/indicators-list/ (accessed February 10, 2021).

[3]UN Department of Economic and Social Affairs: https://sdgs.un.org/2030agenda (accessed February 10, 2021).

[4]UN Habitat has provided a website for urban data: https://unhabitat.org/explore-data (accessed February 10, 2021).

[5]UN Department of Economic and Social Affairs: https://sdgs.un.org/goals (accessed February 10, 2021).

[6]For further details of the Plan, read: https://dilemma-x.net/2014/03/17/china-national-new-type-urbanization-plan-2014-2020/ (accessed February 10, 2021).

While these top-down indicator frameworks provide a very structured monitoring approach to allow comparison across different cities, their aspirational and long-term nature is unlikely to truly reflect the local urban context as drivers of urban development change rapidly and vary widely. More importantly, the development of robust indicator frameworks has to be based on a clear conception of sustainable cities. The development of eco-cities (Lian et al., 2010), sustainable cities (Haughton and Hunter, 1994), smart cities (Giffinger, 2015; Giffinger et al., 2010) and low-carbon cities (Yang and Li, 2013) are among the responses to urban sustainability problems (Alusi et al., 2011). These different concepts, often vaguely defined, coexist and share the principles of sustainable development, but are different in their orientation of desired outcomes and affected by different local contexts and attributes such as economic performance, cultural preference and political environment (Newman and Kenworthy, 1989). Given that an ideal type of sustainable city does not exist (Holden et al., 2008), the logical way forward is to envisage urban sustainability as a development goal rather than confining it in set definitions.

Debates over different concepts and pathways towards urban sustainability highlight significant knowledge gaps in the framing and operationalisation of sustainable development and thus the interest in cities as the intersection of different disciplines. The prospect of using indicators to establish sustainable urban transformation can only be achieved by adopting multi-disciplinary and multi-dimensional approaches to address the conceptual and methodological issues involved (McCormick et al., 2013). The relationship between urbanisation and sustainability under different disciplines in terms of their respective focus, knowledge framing and methodology is highlighted in Table 3.1. Environmental science, for instance, tends to assess environmental quality and impact, economic analysis focuses on modelling the sustainable consumption of natural resources and engineering research of technical solutions, whereas urban planning explores spatial patterns and the environmental consequences associated with them.

Another millstone of the top-down indicator development approach is the highly varied capacity of data infrastructure and local monitoring expertise across different urban contexts, especially cities in developing countries. Since the geographic coverage of SDG indicators at the country level is partial and incomplete (Dunning, 2016), progress monitoring at the city level will be even more problematic. This challenge is clearly evident from the limitations of the UK's top-down online monitoring portal[7] – as five of the fifteen SDG11-related targets have no data associated with them and the other ten targets can only be measured at the UK national level – which is rather ironic as the indicators are meant to monitor sustainability for 'cities and communities'. This suggests that top-down effort alone is inadequate to serve the complex urban agenda. Hence the progress made over the monitoring of longer-term outcome and impact of the planning system has been slow burning across the UK, with most progress made in the devolved administration of Scotland (Scottish Government, 2018) and Wales (Welsh Government, 2019).

---

[7]The UK Office for National Statistics SDGs online reporting platform: https://sustainabledevelopment-uk.github.io/11/ (accessed February 10, 2021).

The discussion here points to the need of having more flexible and adaptable indicator frameworks tailored for different urban contexts and data infrastructure capacities, which goes against the neatly structured, top-down approach introduced by national and supra-national organisations. It is thus unsurprising to see the sprouting of different indicator frameworks and decision toolkits for SDG monitoring at the city level (see Wong and Zheng, 2019). This will encourage cities and their associated communities to use indicators as a collective way of 'seeing' in order to develop their own vision of sustainable urban development (Wong et al., 2015). Examples of bottom-up, participatory indicator frameworks include the sustainability index for Taipei (Lee and Huang, 2007), the compass index of sustainability for Orlando in Florida (Atkisson and Hatcher, 2001) and the UK Urban Future toolkit (Boyko et al., 2012). However, there are also strong arguments for integrating both top-down and bottom-up approaches (Wong and Watkins, 2009), as critical local issues identified from the bottom-up process can be set against the top-down strategic goals and overarching guidance framework. This will allow local communities to articulate their particular concerns and circumstances as well as facilitating governments to implement their strategic goals and plans at the local level.

Table 3.1 Urbanisation and sustainability: different disciplinary perspectives

Research discipline	Research focus	Knowledge framework	Methodology
Environmental science (e.g. Alberti and Marzluff, 2004)	environmental impacts of urbanisation	ecosystem services; green infrastructure; biophysical processes of nature	environmental quality of air, water, vegetation, and biodiversity; eco-footprint
Environmental economics (e.g. Perman et al., 2011)	maximisation of human utility within environmental constraints	green GDP; economics of common-pool resources; valuation of natural resources	cost-benefit analysis; willingness to pay; sustainability indicator
Urban ecology (e.g. Alberti, 2008)	carrying capacity of natural environment and its limits	environmental advocacy; discontinuity in ecological environment	lifestyle change; anti-growth campaign
Urban planning and design (e.g. Jabareen, 2006)	interaction between spatial patterns and urban environment	sustainable urban form; new urbanism; compact city	land-use planning; geographical information system (GIS) analysis; urban growth boundary
Civil engineering and technology (e.g. Neirotti et al., 2014)	technical solutions to environment protection	smart city; eco-city; information and communication technology (ICT); big data; new energy; clean-tech	real-time monitoring; systems of transport; new energy infrastructure
Politics and governance (e.g. Armitage and Plummer, 2010)	cooperation and coordination of actors in protecting environment	multiscale environmental management; governance regime; public participation; organisational strategy	urban environmental policy; congestion tax; pollution/CO_2 permission market

The temporal and multiple-scalar methodological issues

The need to better monitor urban development, with more spatial granularity, is paradoxically compromised by a parallel reduction in the number of robust indicators. Of all the stumbling blocks in indicator research, Wong (2006) argues that 'data' is the crunch point as it is both a requirement and a problem. Without the basic ingredient of high quality datasets, it is simply not possible to produce reliable and robust indicators, though in some cases innovative methodology and analytical techniques can help to ameliorate and overcome some of the problems. The quality threshold of data is particularly important in relation to temporal and multi-scalar methodological issues (Rae and Wong, 2012).

The continuous change in individual communities means that the temporal dynamics of policy concepts such as sustainable development and inclusive growth encapsulate both processes of change and a state of development. Policy formulation and interventions are shaped by specific spatial planning contexts. The indicators used to outline the socio-economic and environmental contexts that spatial planning policies have to interact with and operate under will inevitably overlap with those measuring outcomes (Wong et al., 2006), though any change in contextual indicators is probably attributable to a wide range of factors in which spatial planning may or may not play a part. The fluid relationship between policy contexts and outcomes (Wong and Watkins, 2009) makes it challenging to measure such temporal dynamics to attribute performance to planning policy. It is, therefore, more realistic to view outcomes as the combined effects of change brought about by the planning system and other forces to achieve the objectives of sustainable cities and communities.

Outcomes themselves can be mediated by time and space or the interaction of both. The persistent achievement of planned 'outputs' (e.g. number of social housing) can be used as a proxy measure of immediate effects, which can arguably be defined as planning 'outcomes' (e.g. housing affordability). This then creates a distinction between indicators that are used to ascertain longer-term strategic relevance and those that provide rapid feedback on progress towards policies (Innes and Booher, 2000). Likewise, spatial planning outcomes at one spatial scale may consist of various outputs at another. For instance, urban strategies seeking to facilitate economic growth will require a series of localised outputs in the form of, for example, land being made available and new businesses. Similarly, a series of outputs at the city level may constitute outcomes at the national level. The use of indicators to monitor urban planning is a complex and challenging task and thus requires a more flexible, adaptive approach.

The geographical scale of data and its analysis is critical to the development of meaningful indicators. The variability of indicator values tends to decrease in small areas, whereas intra-area variability tends to increase within larger areas (Openshaw, 1984). This means that indicator values can change markedly with different spatial specifications (Wong, 2015). While sustainable development is a global initiative, it also has tremendous appeal at the local community level. Community, the spatial scale at which land development takes place, is a linchpin of upper-level policy-making and local sustainability delivery (Sharifi and Murayama, 2013; Zheng et al., 2016). Since the issues to be addressed vary at different spatial scales, this points towards the adoption of a multi-scalar approach to monitoring. This provides the flexibility to address different issues with different spatial configurations, but also simultaneously increases the difficulty of monitoring due to the challenge of accessing relevant datasets.

There is also a perpetual tension between the administrative geography and the functional spatial scale for measuring a particular phenomenon; this can be considered a 'scalar mismatch' in policy monitoring. For instance, in the under-bounded city, the administratively defined city is smaller than the physical urban aggregate, while the opposite is true for an over-bounded city (Carter, 1981). The choice of certain boundaries could potentially lead to distorted analysis and misleading conclusions (Wong, 2015). To complicate things further, functional areas (e.g. travel-to-work areas) tend to change over time, possibly in response to the operation of the spatial planning system itself (e.g. decentralisation of housing sites to peri-urban locations).

The objective of conducting meaningful analysis of temporal change and spatial patterns to aid policy design and monitoring could only be accommodated if robust small-area data is available. This explains the reason for a lack of longitudinal study to track the spatial impact of large-scale regeneration projects in the UK. One exception is Schulze Bäing and Wong's (2018) evaluation of the long-term benefits of the flagship regeneration in Salford Quays and its neighbouring areas, which are amongst the most deprived areas in England. The challenge of any such longitudinal study is two-fold: the lack of consistent time-series data and the changing boundaries of the spatial unit of analysis. Their study focuses on examining workplace dynamics and the social sustainability of the area. For workplace analysis, census special workplace statistics were used to identify the changing structure and volume of the workforce in the area, which was followed by an analysis of the commuting flows of the residential locations of the Salford Quays' workforce. The novel geodemographic classification of travel-to-work flows,[8] based on the origin-destination commuting flows developed by the author and others (Hincks et al., 2017), was analysed and mapped to identify where workers with different socio-economic traits commuted (see Figure 3.1). For social sustainability analysis, the official Indices of Multiple Deprivation[9] (Smith et al., 2015) were used to ascertain the spatial distribution and temporal change of deprivation (see Figure 3.2), and HM Land Registry housing transaction data[10] was used to track house price changes in the area.

[8]For the 2011 census of England and Wales, data was collected on commuting between their usual residence and place of work for people aged 16 and over who were in employment in the week before the census. This origin-destination data is referred to as Special Workplace Statistics in which origin-destination flow-data is segmented based on a combination of demographic and socio-economic characteristics. The data consisted of 89 variables covering 11 categories of commuters and was released at Middle Layer Super Output Area (MSOA) level for England and Wales. In 2011, there was 7201 MSOAs (ranging from 5,000 to 15,000 residents) in England and Wales. https://census.ukdataservice.ac.uk/use-data/guides/flow-data.aspx (accessed February 10, 2021).

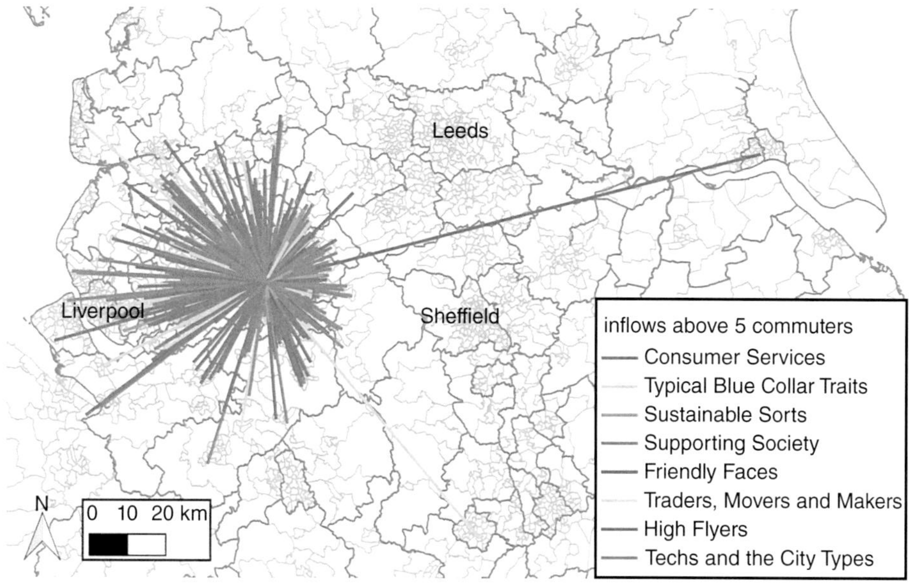

Figure 3.1 Inward commuting flows to Salford Quays by geodemographic classification, 2011

Source: Schulze Bäing and Wong (2018: 524) and www.commute-flow.net

Individual behaviour and its spatial-temporal variation are the main subject and foundation in urban studies and planning practices. Taking commuting flows as an example, individual behaviour of commuting at different spatial-temporal scales can be decomposed into spatial, temporal and individual attributive data. Since the spatial patterns generated are very sensitive to different classification methods, it is the ability to capture different types of data accurately that allows us to develop robust methodology. Despite the rapid development in information-handling capacity since the late 1980s, notably the application of GIS, the affordability of personal computers and the development of spreadsheets, databases and various user-friendly statistical packages, many pressing

[9]The Index of Multiple Deprivation (IMD) is the official measure of relative deprivation for small areas (Lower Super Output Areas with 1,000-3,000 residents) in England, ranking them from 1 (most deprived area) to 32,844 (least deprived area). It consists of seven domains and is combined using different weightings: income deprivation (22.5%); employment deprivation (22.5%); education, skills and training deprivation (13.5%); health deprivation and disability (13.5%); crime (9.3%); barriers to housing and services (9.3%); and living environment deprivation (9.3%). https://www.gov.uk/government/statistics/english-indices-of-deprivation-2019 (accessed February 10, 2021).

[10]https://www.gov.uk/government/organisations/land-registry (accessed February 10, 2021).

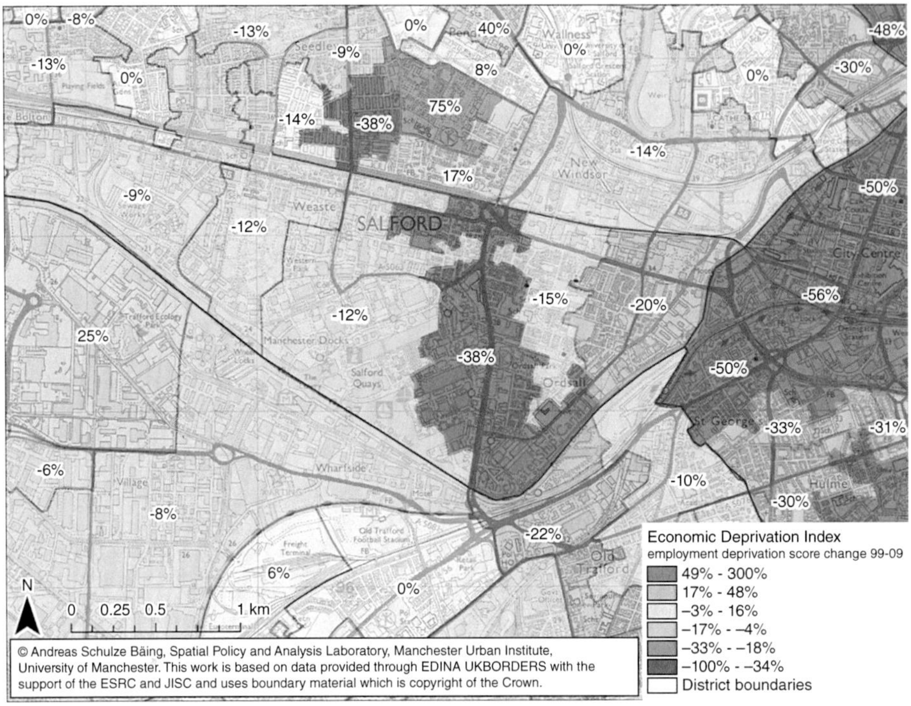

Figure 3.2 Economic Deprivation Index, employment deprivation score change, 1999–2009

Source: Schulze Bäing and Wong (2018: 527)

urban planning issues remain unanswered or only partially answered due to the lack of quality urban data infrastructure (Rae and Wong, 2012; Wong, 2006).

The promise of urban big data: uncertainty and volatility

The prospect of developing a flexible and multi-scalar indicator framework, as discussed above, hinges on access to quality fine-grained data to unlock the potential for urban indicator development. The imperative to develop innovative data analytics comes with the opportunities of unleashing the power of big data for enhanced insight and decision-making. Big data tends to be associated with smart technologies which automatically generate data through sensors of physical objects (e.g. the Internet of Things) in real-time with precise geo-positioning (Batty, 2013). Thus big data is seen as a way of providing 'a more cohesive and smart understanding of the city that enhances efficiency and sustainability' that 'can be used to better depict, model and predict urban processes and simulate the likely outcomes of future' (Kitchin, 2014: 2).

Rather than focusing on the 3Vs (*volume, velocity* and *variety*), Thakuriah et al. (2016: 14–15) define big data as 'structured and unstructured data generated naturally as a part [of] transactional, operational, planning and social activities, or the linkage of such data to purposefully designed data' by focusing on data generation sources and advance a six-fold classification:

1 sensor: infrastructure-based and moving objects;
2 user-generated: participatory sensing systems, web use, social media, etc.;
3 administrative data;
4 private sector: customer and transactions records;
5 arts and humanities: repositories of text, images and recordings, etc.; and
6 hybrid: linked and synthetic data with, for example, survey and census data.

The generation mode of big data (Kitchin, 2014: 4) can be through direction (administrative and surveillance systems, e.g. passport control), automation (inherent automatic function of devices and systems) and volunteering (crowdsourcing). However, it is automated data that has caught most attention and is seen as instrumental in shaping smart solutions and the smart planning of our cities. Since big data is more effectively organised in a network structure rather than in a traditional tree structure, Pan et al. (2016) argue that its classification should be made on the basis of the data processing method and the application objectives.

By taking a more critical stance to examine the nature of big data, Kitchin (2014) highlights the fact that the framing, classification and use of big data, just like other data, is socially situated and contextually-based and subject to the constraints of technical know-how, ethical considerations and public opinion, as well as the regulatory environment, funding and available resources. Shearmur (2015) further points out the paradox of big data – despite its 'big' size, its scope is indeed very narrow. This is because big data tends to be passively generated by users and market-driven, which creates inherent bias in their focus and will exclude those who are not engaged in particular activities or markets. As passive big data does not have a strong focus on 'society', the data streams do not include the attributes of individuals (e.g. age, occupation, qualification, and neighbourhood) to allow more nuanced analysis.

The strengths of big data are also its weaknesses. Its automatic and spontaneous nature means that it is not structured or framed by concepts and definitions that are most relevant to address research needs and policy-making. While big data can be exploited opportunistically to support research, data-rich approaches are often theoretically poor (Shearmur, 2015). The sheer volume of diverse data that has emerged at such a quick speed also draws our attention to the timeliness of information, which can encourage short-termism and place more emphasis on the moment. Batty (2013) thus argues that new theories that are able to deal with diverse time horizons are needed to underpin the framing of research analysis to avoid the trap of short-termism. The challenge is succinctly articulated by West (2013):

> The trouble is, we don't have a unified, conceptual framework for addressing questions of complexity. We don't know what kind of data we need, nor how much, or what critical questions we should be asking. 'Big data' without a 'big theory' to go with it loses much of its potency and usefulness, potentially generating new unintended consequences.

Unlocking the potential of big data: opportunities and challenges

As stated on Oracle's website,[11] 'Finding value in big data isn't only about analyzing it (which is a whole other benefit). It's an entire discovery process that requires insightful analysts, business users, and executives who ask the right questions, recognise patterns, make informed assumptions, and predict behaviour.' This echoes the concern voiced by Tukey (1986) that data alone cannot provide an answer; it is very much about problem-framing as well as insightful analysis. As big data is user- and market-oriented, its application in the urban planning field is even more challenging.

The main value of big data is to address the long-standing, temporal-spatial methodological issues of measurement as these new data sources, coupled with the characteristics of high frequencies of change and refined geo-referencing, open the possibilities to examine urban questions that were previously found impenetrable (Glaeser et al., 2018; Pan et al., 2016). In other words, by integrating, processing, analysing and visualising large amounts of both structured and unstructured data from a unified urban data platform, new opportunities for the development of indicators will be opened up. Despite these promises, the development of smart city analytics is still at the early stage (Anejionu et al., 2019). The application of big data tends to be mostly found in the environmental, transport and infrastructure planning fields (e.g. Anda et al., 2017; Bibri, 2018; Cottrill and Derrible, 2015; Hao et al., 2015; Rathore et al., 2016).

Some of these nascent attempts include the establishment of the UK Urban Big Data Centre[12] in 2014 with funding from the Economic and Social Research Council. One of the Centre's key activities is to develop a Spatial Urban Data System (SUDS) as a spatial big data infrastructure that covers a wide range of social, economic, natural- and built-environment and physical infrastructure indicators for fourteen built-up areas[13] through using novel forms of data, modelling and simulation programs. As explained by Anejionu et al. (2019), the urban indicators generated by SUDS differ from other similar systems as they are regulated and updated by spatially activated metrics from (near-) real-time data sources.

Another example is the use of Google Street View by Glaeser et al. (2018), via a computer-based visual recognition technique of street images, to develop an income prediction

---

[11] https://www.oracle.com/uk/big-data/guide/what-is-big-data.html (accessed February 10, 2021).

[12] https://www.ubdc.ac.uk/ (accessed February 10, 2021).

model across New York City block groups. They argue that by combining big data sources such as imagery with predictive models, the gaps in a variety of smaller-sample data sources can be bridged. However, their prediction model results in a Pearson's correlation coefficient R^2 of 0.77 (i.e. explaining 59.3% of variation) when used to predict income out-of-sample in New York City and a R^2 of 0.71 (i.e. explaining 50.4% of variation) in Boston. The question is whether the prediction of less than 60% is acceptable. Hence, Glaeser et al. (2018) emphasise that the use of imagery data and prediction methods is particularly beneficial for mapping wealth and poverty in the developing world, where reliable data and measurement are lacking. Both of these examples illustrate an obvious strength of big data in terms of using images and other information to develop urban indicators as well as presenting their values and patterns through geo-visualisation mapping.

While big data opens up new avenues for analysing and presenting spatial indicator values, especially dynamic flow analysis, it is less useful in terms of supplying attribute data for tabulations and data linkage due to strict privacy, data protection and ethics regulation environments (Kitchin, 2013). This means that big data tends to focus on spatial patterns and cannot be related to attributes to develop more nuanced analyses such as those shown in Figure 3.1. The only way forward to link attribute data with count/flow data will involve census or survey data. For example, Hincks et al. (2017) performed a geo-demographic classification of commuting flows data with 2011 Census Special Workplace data (with an associated online mapping portal www.commute-flow.net) and Zheng et al. (2019) classified commuting flows in Beijing suburban neighbourhoods by different socio-economic groups based on their own survey data.

Apart from technical capacities and skills, the fundamental challenge to the use of big data rests with its 'passive' nature. Since big data sources are generated through sensing technologies and user crowdsourcing, their coverage and representativeness vary and are incomplete, which increase their complexity and volatility. In methodological terms, the incomplete coverage of any big data sources can be regarded as selective samples which inevitably involve bias and errors in the measurement and estimation of key parameters (Kitchin, 2013; Shearmur, 2015). Since the knowledge of full population data is not necessarily available, it would be difficult to accurately estimate the bias and errors involved in such big data sources.

The strong potential of using urban big data lies in its ability to link, integrate and connect a large variety of data sources, but reality often falls short of ideal, which is illustrated in the study of London transport data by Batty (2013). The large amount of travel data generated from London's smart 'Oyster' card was undermined by the fact that only 85% of travellers used the card, which tends to exclude occasional users and tourists. Even though the data

---

[13]Examples of the indicators include percentage of persons in full employment (%); unemployed: aged 16-24 years (%); unemployment rate: all economically active only (%); public green space; and job accessibility indicators for Bristol, Birmingham, Cardiff, Leeds, Liverpool, London, Manchester, Nottingham, Sheffield, Southampton-Portsmouth, Glasgow, Edinburgh, Belfast. https://www.ubdc.ac.uk/data-services/data-catalogue/geospatial-data/suds-spatial-urban-data-system-urban-indicators/ (accessed February 10, 2021).

covers tap-in and tap-out counts, they cannot be treated as origins and destinations of trips to feed into traditional transport models. Furthermore, the linkage between the Oyster card data and the geo-positioning data of trains and buses from Transport for London is hindered because there is no way to find out which individual passenger gets onto which train or bus. That is precisely the reason why Shearmur (2015) strongly argued against any attempt to replace census-type data with the potential grand promise of big data.

With these challenges to the use of urban big data in measurement, a consensus has emerged among academics that while big data has opened up the possibilities and new resources to improve our research and understanding of cities, there is a need to pay more heed to its usage and be mindful of its limitations and potential errors. Therefore, the recommendation is that big data is best when used in conjunction with other independent data sources (see Batty, 2013; Glaeser et al., 2018; Shearmur, 2015) to develop robust analysis and intelligence to inform policy-making. By focusing on the development of urban indicators, it is pertinent to ask what kind of indicator frameworks should be formulated under the strong wave of urban big data. A holistic methodological approach was developed to combine different types of indicators to form a strategic framework for measuring spatial planning outcomes by Wong and Watkins (2009). This framework moves further towards the development of an integrative and communicative framework for indicator development, defined by four core principles (Rae and Wong, 2012: 884–8). It is useful to revisit these principles in the face of urban big data:

1. 'Indicators are parts of a whole: Individual indicators should not be used separately, but within an integrative framework which views them as parts of a whole. A coherent group of indicators can then be identified within each theme or domain (e.g. inclusive and liveable communities, sustainable economic development) to reflect the objectives of any spatial plan.'

   Given that big data consists of a mix of structured and unstructured data organised in a network structure, rather than in a neat hierarchical form, the need to treat indicators as parts of an integrative whole through complementary usage is even more pertinent in establishing a robust picture of urban issues.
2. 'Combine indicators across themes: The strategic bundling method places emphasis on combining multiple indicators across different themes, and includes different types of indicator (e.g. contextual, process, outcome) in the analysis. This approach involves using indicators in combination but not producing a single synthesis or summary value.'

   Since the intrinsic value of big data is its power of integrating and connecting a large number of data from different sources in an automatic and frequent manner, it should theoretically provide the flexibility and scope to innovatively develop and combine different measures to perform cross-cutting thematic analysis. However, the stumbling block to combining indicators lies in the differential sampling coverage and representativeness of data from an asymmetrical data structure.
3. 'Stakeholder ownership: Each spatial planning region (e.g. city-region, metropolitan area) should take a lead in developing its spatial planning indicator framework. By focusing on a common vision and collective outcomes within each region, the framework should become a shared learning experience between key local and regional stakeholders.'

In spite of the availability of big data sources, their access, processing, storage and analysis do pose major challenges as they require both technical and policy knowledge. Since many of these datasets are either held by the market/organisation or crowdsourced through social media, a collaborative approach to the development of indicators, governed by robust privacy and data protection regulation, is thus becoming critical, as no one single group of stakeholders can address the gravity of these challenges to reap the benefits.

4 'A spatial context for indicators: The final principle is to view individual areas within a wider spatial context rather than as isolated locations. Stakeholders ought to think about critical development issues in different parts of their region and exercise flexibility to choose any extra indicators relevant to their specific spatial context.'

With the advance of urban big data and smart urbanism, often associated with sensor data and innovative visualisation methods, indicator analysis of different spatial contexts at different spatial scales should play to the strengths of big data.

Conclusion

This chapter provides a critical discussion over the opportunities and challenges brought by big data to unlock the potential of urban indicator development. The discussion points to the fact that we have to go beyond the euphoric veil of big data to focus on the problems in hand and be realistic about what big data can offer in the near future. The foremost challenge is having quality data for the development of robust indicators. Other than the 3Vs, Kitchin (2014: 3) highlights four more features of big data – being exhaustive in scope; fine-grained in resolution; relational in nature; and being flexible – which indeed help to qualify what constitutes robust urban big data.

When encountering an overwhelming amount of data, we need to figure out how best to make sense of the fog of information sources by processing, evaluating and analysing them. As shown by the British experience, information overload can just create irritating noise in policy monitoring and interfere with our ability to learn and engage in creative problem-solving (Wong, 2006). In a way, the digital revolution and big data are both a blessing and a curse, as the processing and analysis of a large amount of data from different formats forms a black box that requires a lot of technical know-how and imagination to handle. However, like the previous experience with GIS, it is problem-framing and policy knowledge that help to extract relevant intelligence for policy-making (e.g. Gilfoyle and Wong, 1998; Worrall, 1994). This means that we have to train a new generation of spatial planners who have a strong grip of policy issues as well as being technically astute.

Finally, the international momentum of pushing a new urban agenda places a concomitant demand for the development of indicators to monitor progress towards urban sustainability. Big data is clearly a valuable resource and arguably can provide the raw material for spatial indicator development in places where there is no established data infrastructure to monitor progress towards SDGs. Given that many places in the developing world start policy monitoring from a blank sheet, big data could be a pragmatic way forward for developing more bottom-up, adaptable indicator frameworks for policy monitoring. More importantly, in the short term, the experimental use of big data can encourage communities to use indicators as a collective way of 'seeing' to develop their own data infrastructure (Wong et al., 2015). Although the community's own vision may be at odds with the strategic goals of sustainable development, it will allow them to engage with the government to align their expectation of sustainable development. Nevertheless, concerted effort across different stakeholder groups will be required in the medium and long term to ameliorate the identified shortcoming of big data before exploiting the full advantage of such data sources.

Acknowledgements

Some of the materials in this chapter are based on findings from two projects funded by the UK Economic and Social Research Council (Grant Numbers ES/N010698/1 and ES/L014459/1). The author would like to thank Dr. Miao Qiao for assisting the compilation of Table 3.1 and Dr. Andreas Schulze Bäing for producing the commuting flow and deprivation maps.

References

Alberti, M. (2008) *Advances in Urban Ecology: Integrating Humans and Ecological Processes in Urban Ecosystems*. Seattle, WA: Springer.

Alberti, M. and Marzluff, J.M. (2004) 'Ecological resilience in urban ecosystems: Linking urban patterns to human and ecological functions', *Urban Ecosystems*, 7: 241–65.

Alusi, A., Eccles, R.G., Edmondon, A.C. and Zuzul, T. (2011) 'Sustainable cities: Oxymoron or the shape of the future?', *Harvard Business School Working Paper*, www.hbs.edu/faculty/Publication%20Files/11-062.pdf (accessed February 10, 2021).

Anda, C., Erath, A. and Fourie, P.J. (2017) 'Transport modelling in the age of big data', *International Journal of Urban Sciences*, 21: 19–42.

Anejionu, O.C.D., Thakuriah, P., McHugh, A., Sun, Y., McArthur, D., Mason, P. and Walpole, R. (2019) 'Spatial urban data system: A cloud-enabled big data infrastructure for social and economic urban analytics', *Future Generation Computer Systems*, 98: 456–73.

Armitage, D. and Plummer, R. (eds) (2010) *Adaptive Capacity and Environmental Governance*. Berlin: Springer.

Atkisson, A. and Hatcher, R.L. (2001) 'The compass index of sustainability: Prototype for a comprehensive sustainability information system', *Journal of Environmental Assessment Policy and Management*, 3 (4): 509–32.

Batty, M. (2013) 'Big data, smart cities and city planning', *Dialogues in Human Geography*, 3 (3): 274–9.

Bibri, S.E. (2018) 'The IoT for smart sustainable cities of the future: An analytical framework for sensor-based big data applications for environmental sustainability', *Sustainable Cities and Society*, 38: 230–53.

Boyko, C.T., Gaterell, M.R., Barber, A.R., Brown, J., Bryson, J.R., Butler, D., Caputo, S., Caserio, M., Coles, R., Cooper, R. and Davies, G. (2012) 'Benchmarking sustainability in cities: The role of indicators and future scenarios', *Global Environmental Change*, 22 (1): 245–54.

Carter, H. (1981) *The Study of Urban Geography* (3rd edition). London: Edward Arnold.
China State Council (2014) 国家新型城镇化规划 (2014–2020年) *[National New Urbanisation Plan (2014–2020)]*. www.gov.cn/zhengce/2014-03/16/content_2640075.htm (accessed February 10, 2021).
Cottrill, C.D. and Derrible, S. (2015) 'Leveraging big data for the development of transport sustainability indicators', *Journal of Urban Technology*, 22 (1): 45–64.
Dunning, C. (2016) '230 indicators approved for SDG Agenda', *Center for Global Development*, 15 March. www.cgdev.org/blog/230-indicators-approved-sdg-agenda (accessed February 10, 2021).
Giffinger, R. (2015) 'Smart city concepts: Chances and risks of energy efficient urban development', in M. Helfert, K.-H. Krempels, C. Klein, B. Donnellan and O. Gusikhin (eds), *Smart Cities, Green Technologies, and Intelligent Transport Systems*. Basel: Springer. pp. 3–16.
Giffinger, R., Haindlmaier, G. and Kramar, H. (2010) 'The role of rankings in growing city competition', *Urban Research & Practice*, 3 (3): 299–312.
Gilfoyle, I. and Wong, C. (1998) 'Computer applications in planning: Twenty years' experience of Cheshire County Council', *Planning Practice and Research*, 13 (2): 191–7.
Glaeser, E.L., Kominers, S.D., Luca, M. and Naik, N. (2018) 'Big data and big cities: The promises and limitations of improved measures of urban life', *Economic Inquiry*, 56 (1): 114–37.
Hao, J., Zhu, J. and Zhong, R. (2015) 'The rise of big data on urban studies and planning practices in China: Review and open research issues', *Journal of Urban Management*, 4: 92–124.
Haughton, G. and Hunter, C. (1994) *Sustainable Cities*. London: Jessica Kingsley.
Hincks, S., Kingston, R., Webb, B. and Wong, C. (2017) 'A new geodemographic classification of commuting flows for England and Wales', *International Journal of Geographical Information Science*, 32 (4): 663–84.
Holden, M., Roseland, M., Ferguson, K. and Perl, A. (2008) 'Seeking urban sustainability on the world stage', *Habitat International*, 32 (3): 305–17.
Innes, J.E. and Booher, D.E. (2000) 'Indicators for sustainable communities: A strategy for building on complexity theory and distributed intelligence', *Planning Theory and Practice*, 1 (2): 173–86.
Jabareen, Y.R. (2006) 'Sustainable urban forms: Their typologies, models and concepts', *Journal of Planning Education and Research*, 26: 38–52.
Kitchin, R. (2013) 'Big data and human geography: Opportunities, challenges and risks', *Dialogues in Human Geography*, 3 (3): 262–7.
Kitchin, R. (2014) 'The real-time city? Big data and smart urbanism', *GeoJournal*, 79: 1–14.
Laney, D. (2001) '3D data management: Controlling data volume, velocity, and variety, application delivery strategies', *META Group*, February 6. https://web.archive.org/web/20120304154148/https://blogs.gartner.com/doug-laney/files/2012/01/ad949-3D-Data-Management-Controlling-Data-Volume-Velocity-and-Variety.pdf (accessed February 10, 2021).
Laney, D. (2012) 'Deja VVVu: Others claiming Gartner's construct for big data', *Gartner Blog Network*, 14 January. https://web.archive.org/web/20120120125408/http://blogs.gartner.com/doug-laney/deja-vvvue-others-claiming-gartners-volume-velocity-variety-construct-for-big-data/ (accessed February 10, 2021).
Lee, Y.J. and Huang, C.M. (2007) 'Sustainability index for Taipei', *Environmental Impact Assessment Review*, 27 (6): 505–21.
Lian, K.K., Gunawansa, A. and Bhullar, L. (2010) '"Eco-cities" and "sustainable cities" – whither?', *Social Space, Lien Centre for Social Innovation*, 3: 84–92.

Manyika, J., Chui, M., Brown, B., Bughin, J., Dobbs, R., Roxburgh, C. and Byers, A.H. (2011) *Big Data: The Next Frontier for Innovation, Competition, and Productivity*. New York, NY: McKinsey Global Institute.

McCormick, K., Anderberg, S., Coenen, L. and Neij, L. (2013) 'Advancing sustainable urban transformation', *Journal of Cleaner Production*, 50 (1): 1–11.

Neirotti, P., Marco, A.D., Cagliano, A.C., Mangano, G. and Scorrano, F. (2014) 'Current trends in smart city initiatives: Some stylised facts', *Cities*, 38: 25–36.

Newman, P.G. and Kenworthy, J.R. (1989) *Cities and Automobile Dependence: An International Sourcebook*. Aldershot: Gower.

Openshaw, S. (1984) *The Modifiable Areal Unit Problem*. Norwich: GeoBooks.

Pan, Y., Tian, Y., Liu, X., Gu, D. and Hua, G. (2016) 'Urban big data and the development of city intelligence', *Engineering*, 2: 171–8.

Perman, R., Ma, Y., Common, M., Maddison, D. and McGilvray, J. (2011) *Natural Resource and Environmental Economics* (4th edition). Harlow: Pearson Education.

Press, G. (2014) '12 big data definitions: What's yours?', *Forbes online*, 3 September. https://www.forbes.com/sites/gilpress/2014/09/03/12-big-data-definitions-whats-yours/#429d823913ae (accessed February 10, 2021).

Rae, A. and Wong, C. (2012) 'Monitoring spatial planning policies: Towards an analytical, adaptive and spatial approach to a "wicked problem"', *Environment and Planning B: Planning and Design*, 39: 880–96.

Rathore, M.M., Ahmad, A., Paul, A. and Rho, S. (2016) 'Urban planning and building smart cities based on the Internet of Things using big data analytics', *Computer Networks*, 101: 63–80.

Schulze Bäing, A. and Wong, C. (2018) 'The impact of brownfield regeneration on neighbourhood dynamics: The case of Salford Quays in England', *Town Planning Review*, 89 (5): 513–34.

Scottish Government (2018) *Monitoring the Outcomes of Planning: A Research Study*. Scottish Government, Edinburgh. Available from https://www.gov.scot/publications/monitoring-outcomes-planing-research-study/ (accessed February 10, 2021).

Sharifi, A. and Murayama, A. (2013) 'A critical review of seven selected neighborhood sustainability assessment tools', *Environmental Impact Assessment Review*, 38: 73–87.

Shearmur, R. (2015) 'Dazzled by data: Big data, the census and urban geography', *Urban Geography*, 36 (7): 965–8.

Sicular, S. (2013) 'Gartner's big data definition consists of three parts, not to be confused with three "V"s', *Forbes online*, 27 March. https://www.forbes.com/sites/gartnergroup/2013/03/27/gartners-big-data-definition-consists-of-three-parts-not-to-be-confused-with-three-vs/#6317dcb942f6 (accessed February 10, 2021).

Smith, T., Noble, M., Noble, S., Wright, G., Mclennan, D. and Plunkett, E. (2015) *The English Indices of Deprivation 2015* (Research report). London: Department for Communities and Local Government. Available from https://www.gov.uk/government/statistics/english-indices-of-deprivation-2015 (accessed February 10, 2021).

Sustainable Cities International (2012) *Indicators for Sustainability: How Cities Are Monitoring and Evaluating Their Success*. Vancouver: Canadian International Development Agency.

Thakuriah, P., Tilahun, N. and Zellner, M. (2016) 'Big data and urban informatics: Innovations and challenges to urban planning and knowledge discovery', in P. Thakuriah, N. Tilahun and M. Zellner (eds), *Seeing Cities Through Big Data: Research Methods and Applications in Urban Informatics*. New York, NY: Springer. pp. 11–45.

Tukey, J. (1986) 'Sunset Salvo', *The American Statistician*, 40 (1): 72–6.

United Nations (2015) *Transforming Our World: The 2030 Agenda for Sustainable Development A/RES/70/1*. https://sustainabledevelopment.un.org/content/documents/21252030%20Agenda%20for%20Sustainable%20Development%20web.pdf (accessed February 10, 2021).

United Nations (2016) *New Urban Agenda* (Habitat III, Quito). United Nations Conference on Housing and Sustainable Urban Development. https://uploads.habitat3.org/hb3/NUA-English.pdf (accessed February 10, 2021).

United Nations Commission on Sustainable Development (UNCSD) (1996) *Indicators of Sustainable Development: Framework and Methodologies*. New York: UNCSD.

United Nations Department of Economics and Social Affairs (UNDESA) (2012) *World Urbanization Prospects – The 2011 Revision*. New York: UNDESA. https://www.un.org/en/development/desa/population/publications/pdf/urbanization/WUP2011_Report.pdf (accessed February 10, 2021).

United Nations Department of Economics and Social Affairs (UNDESA) (2017) *Inter-agency and Expert Group on SDG Indicators*. New York, NY: UNDESA. https://unstats.un.org/sdgs/iaeg-sdgs/ (accessed February 10, 2021).

Welsh Government (2019) *Well-being of Wales 2018–2019*. Cardiff: Welsh Government. https://gov.wales/sites/default/files/statistics-and-research/2019-09/well-being-of-wales-2019-590.pdf (accessed February 10, 2021).

West, G. (2013) 'Big data needs a big theory to go with it', *Scientific American*, 1 May. https://www.scientificamerican.com/article/big-data-needs-big-theory/ (accessed February 10, 2021).

Wong, C. (2006) *Quantitative Indicators for Urban and Regional Planning: The Interplay of Policy and Methods*. London: Routledge.

Wong, C. (2015) 'A framework for "City Prosperity Index": Linking indicators, analysis and policy', *Habitat International*, 45: 3–9.

Wong, C. and Watkins, C. (2009) 'Conceptualising spatial planning outcomes: Towards an integrative measurement framework', *Town Planning Review*, 80 (4/5): 481–516.

Wong, C. and Zheng, W. (2019) 'Indicator-based approaches to environmental planning', in S. Davoudi, R. Cowell, I. White and H. Blanco (eds), *The Routledge Companion to Environmental Planning and Sustainability*. London: Routledge. pp. 329–45.

Wong, C., Baker, M. and Kidd, S. (2006) 'Monitoring of spatial strategies: The case of *local development documents* in England', *Environment and Planning C: Government and Policy*, 24 (4): 533–52.

Wong, C., Baker, M., Webb, B., Hincks, S. and Schulze Bäing, A. (2015) 'Mapping policies and programmes: The use of GIS to communicate spatial relationships in England', *Environment and Planning B: Planning and Design*, 42 (6): 1020–39.

Wong, C., Qiao, M. and Zheng, W. (2018) 'Dispersing, regulating and upgrading urban villages in suburban Beijing', *Town Planning Review*, 89 (6): 597–621.

Worrall, L. (1994) 'The role of GIS-based spatial analysis in strategic management in local government', *Computer, Environment and Urban Systems*, 18 (5): 323–32.

Yang, L. and Li, Y. (2013) 'Low-carbon city in China', *Sustainable Cities and Society*, 9: 62–6.

Zheng, H.W., Shen, G.Q., Song, Y., Sun, B. and Hong, J. (2016) 'Neighborhood sustainability in urban renewal: An assessment framework', *Environment and Planning B: Urban Analytics and City Science*, 44 (5): 903–24.

Zheng, W., Wong, C. and Qiao, M. (2019) 'Socio-spatial variations in commuting patterns in suburban Beijing', *Built Environment*, 45 (4): 523–43.

4

CROWDSOURCED GEOGRAPHIC INFORMATION FOR URBAN PLANNING AND MANAGEMENT

Adipandang Yudono

Introduction

The aim of this chapter is to demonstrate how crowdsourced geographic information can form the basis of urban planning and management analysis. It explores Indonesian geospatial policies and regulations through a case study using archival research and in-depth interviews with city councils and OSM (OpenStreetMap) organisers. A complex societal system needs citizen participation to maintain its living harmony with social and environmental change, including urban planning and management programmes initiated by governments (Smith, 1973). The rationale for citizen participation in every planning programme is even more important in enabling individuals and communities to engage more intimately with environmental modification. With citizen participation, information and judgement pertaining to local systems can provide greater proximity and accuracy to a planning process (Smith, 1973).

Citizen participation in crowdsourced geographic information can be defined as a political process to achieve planning and management consensus between diverse interests by providing public opportunities for dialogue with city councils and local government representatives, as well as planners and the community itself. Citizen participation methods can be practised in various ways: such as individual voluntary digitising of geographic data, mapathons, participatory mapping, public hearings, community surveys, advisory groups and committees, focus groups, community ombudsman/complaint centres.

In Indonesia, citizen participation in crowdsourced geographic information for urban planning and management is carried out in the form of the presentation of information, suggestions and advice orally and in writing through a variety of information media. In the context of the delivery of information from citizens to government agencies, maps are usually used as the negotiation medium to achieve planning and management consensus. Negotiating through maps makes a key contribution to achieving spatial planning and management consensus. In consideration of this, the focus of this discussion is to explore and examine how crowdsourced geographic information practices in Indonesia provide large amounts of detailed geographic information as an essential factor for urban planning and management analysis.

Underlying the idea of crowdsourced geographic information

Interactive internet usage is growing rapidly with the next-generation of Web 3.0 which allows not only the interaction between two or more internet users and information providers, but also the sharing of files. These developments, along with facilities for data and information sharing, have changed the outlook on the production and use of data, since now the user can also be involved as a data and information producer and provider (see Figure 4.1).

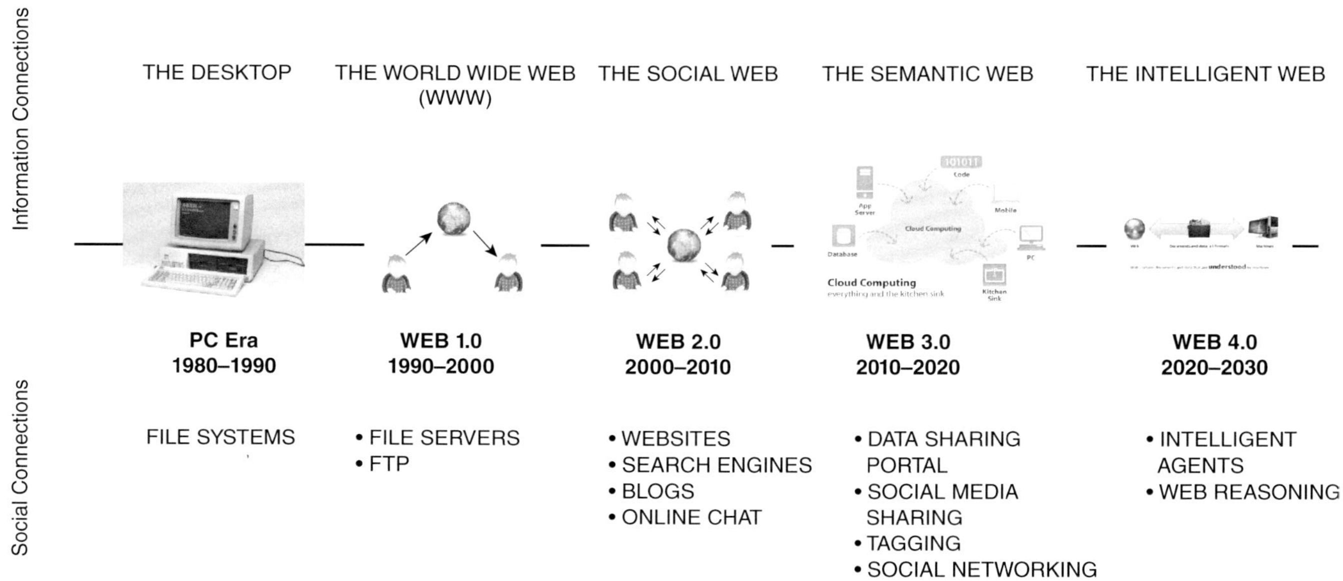

Figure 4.1 Internet generation era

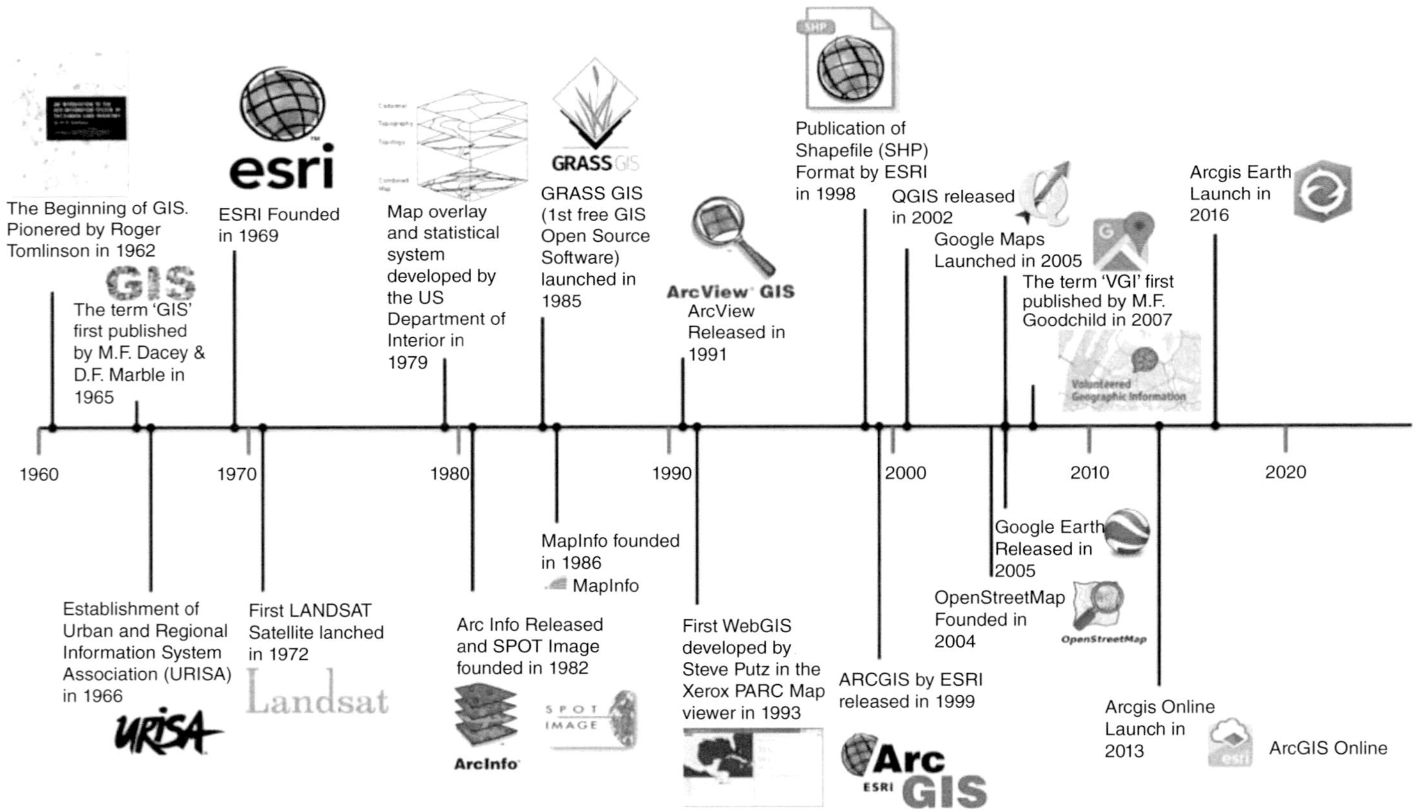

Figure 4.2 GIS timeline

In terms of spatial data sharing in the Web 3.0 context, citizens can create and share geographic data and information via public web mapping application programming interfaces (APIs) such as Google Maps, Wikimapia, OSM and Microsoft's Virtual Earth (Goodchild, 2007; Rouse et al., 2007; Tulloch, 2007). Historically, in the 1990s, geographic data production and provision was dominated by national mapping agencies. However, since the 2000s, geographic data providers outside government agencies have been widely established to produce and provide geographic data development and sharing services amongst their communities through greater support of the internet and mapping technology advances that provide open-source GIS for the production and provision of geographic data and information. The development of GIS from a stand-alone application to crowdsourced geographic information can be seen in Figure 4.2.

Crowdsourced geographic information is emerging rapidly in most areas of the world. In the past, only geographers or cartographers could create accurate geospatial data. Now, through the use of supporting advanced devices, such as the Global Positioning System (GPS), smartphones and cloud computing, geospatial data can be produced, collected, stored, disseminated, analysed, visualised and used by people with no geography or cartography background (Sui et al., 2013). Furthermore, Gould (1999) argues that, in this digital era, the advances in technology and the multiple devices available that can be equipped with sensors for creating geographic features, mean that everybody can be a geographer.

Yet there has been little consideration of how geographic information sharing is practised in civil societies. Since open-source participatory geographical information systems (GIS) platforms were released, such as OpenStreetMap (OSM) in 2004, a person or group of people with no cartographic skills and working outside government agencies can create and produce geographic data voluntarily (Sui et al., 2013). In terms of urban management purposes, geographic data or information produced by local communities may help to create local spatial knowledge to support spatial policymaking.

Crowdsourced geographic information has the possibility to be a useful democratic planning tool to empower citizens participating in geographic data provision and sharing in government programmes (Johnson and Sieber, 2013). Since the global adaptation of World Wide Web (WWW) technology, the interaction between information providers and users has increased. Local communities are able to produce and share geographic data to produce web interfaces with territorial information in public mapping application programming interfaces (APIs) such as Google Maps, OSM and Wikimapia (Goodchild 2007; Rouse et al., 2007; Tulloch, 2007).

In terms of democratic spatial planning action, crowdsourced geographic information is considered an effective voluntary method of helping people feel comfortable with the required technology and other co-participants in order to shape coalitions of local knowledge (Craig and Elwood, 1998). However, this voluntary behaviour by crowdsourced geographic information communities raises critical questions about individual motivations for contributing to geographic data production and sharing. Coleman et al. (2009) identified the following eleven aspects of the volunteered (crowdsourced) geographic information (VGI):

1 Altruism;
2 Professional or personal interest;
3 Intellectual stimulation;
4 Protection or enhancement of a personal investment;
5 Social reward;
6 Enhanced personal reputation;
7 Outlet for creative & independent self-expression;
8 Pride of place.

And, on the negative side:

1 Mischief;
2 Social, economic or political agenda; and
3 Malice and/or criminal intent.

Source: Coleman et al. (2009: 343–5)

Crowdsourced geographic information uses sharing and exchange of geographic information activities to interact with other stakeholders to create spatial narratives (DiBiase et al., 1992). Within this perspective, crowdsourced geographic information may contribute to the creation of local spatial knowledge by examining, synthesising and formulating geographic data and information related to communities' neighbourhood issues (Elwood, 2006). However, even though crowdsourced geographic information is a useful approach to empower local stakeholders in terms of urban planning and management, crowdsourced geographic information participants from different professions, ages and education levels may degrade data quality performance; for instance, data may be less accurate or use non-standardised geo-referenced systems.

Assessment of the quality of crowdsourced geographic information

Since 2002, the quality of geographic information has been regulated by the International Organization for Standardization (ISO): Code 19113 examines quality principles and Code 19114 considers the quality of evaluation procedures. Both ISO codes are organised under the aegis of the ISO Technical Committee 211 (Haklay, 2010). Today, both ISO catalogues have been revised and merged into one ISO catalogue – ISO 19157:2013, Geographic Information – data quality. The quality standards of geographic information assessment based on ISO 19157:2013 have been categorised into six groups: completeness, thematic accuracy, logical consistency, temporal quality, usability element, and positional accuracy.

The assessment of the quality of crowdsourced geographic information generates debate but, as yet, there is no consensus about mechanisms or methods (Antoniou and

Skopeliti, 2015). However, many geography scholars have tried to measure the quality of crowdsourced geographic information in their own approaches using ISO standard parameters. For instance, Haklay (2010) by comparing the OSM datasets and Ordnance Survey datasets; UGM and the Indonesian HOT (2012) by comparing the OSM datasets with reference datasets taken from field surveys; Antoniou and Skopeliti (2015) by reviewing comprehensive VGI literatures with ISO 19157:2013 standard; Basiri et al. (2016) by finding bugs and errors based on spatial knowledge of anonymous tracking data users; Senaratne et al. (2016) by proposing data mining; Fonte et al. (2017) by reviewing VGI datasets (OSM and Wikimapia) with ISO 19157:2013 standard. Coleman et al. (2009) generally state that the quality of crowdsourced geographic information is varied, but the approaches reviewed by Antoniou and Skopeliti (2015) and Fonte et al. (2017) are convergent to global geographic data quality assessment. Crowdsourced geographic information established by international organisation reputation, such as OpenStreetMap (OSM), Wikimapia and Google Map Maker has greater accuracy (with a margin of error of less than 10%) than crowdsourced geographic information established under local reputation. Overall, even though spatial data inputting to open-source GIS platforms is conducted by ordinary mapping contributors, it is suggested that crowdsourced geographic information organisers and administrators should have geography, geodetic or cartography backgrounds in order to edit, manipulate and revise erroneous geographic features to be more accurate.

The value of crowdsourced geographic information

The value of crowdsourced geographic information can be evaluated from two perspectives: the participants of the crowdsourced geographic information community; and the geographic information that is produced, provided, disseminated and shared amongst communities and other stakeholders, such as governments and planning professions.

With regard to the participants of crowdsourced geographic information groups, the main value of crowdsourced geographic information can be identified as contributing to a democratic system. Crowdsourced geographic information works by empowering citizens to create geographic information, which will provide geographic data to improve government policymaking. Crowdsourced geographic information can:

1. Foster a bottom-up participation approach in planning processes by allowing citizens control over their own living places (Feick and Roche, 2013).
2. Improve social networks and individual technical skills among crowdsourced geographic information community members (Feick and Roche, 2013).
3. Share data amongst local crowdsourced geographic information communities to create local spatial knowledge. Local communities can understand and manage their own areas; better data and information created by local communities can provide evidence and tools for negotiation with government over planning and development agendas (Obermeyer, 2006; Roche and Caron, 2009).

The value of crowdsourced geographic information for geographic data or information production can be listed as follows:

1 Geographic data or information that is produced, provided, stored and disseminated by crowdsourced geographic information communities will have primarily latent value when used for an applicable issue and cognitive task (Feick and Roche, 2013). A single type of data might be used in various techniques, sometimes when mixed with other data, which are then jointly referenced or analysed in unique methods, combining data and producing new information (Longhorn and Blakemore, 2008).
2 Information is a non-physical product that frequently relates to a *de facto* public good since use by one individual does not inhibit others' use; but usage limitations due to licence agreements, copyright and confidentiality concerns sometimes imply that it can approximate a *de jure* private good (Feick and Roche, 2013).
3 With a large number of crowdsourced geographic information participants, the geographic data or information created by crowdsourced geographic information communities may be far more responsive to creating local spatial knowledge than that prepared by governments or private firms, which may lack the necessary budgets, resources and time. For instance, when a disaster occurs in a particular area, OpenStreetMap may permit quick crisis responses by involving the global crowdsourced geographic information community to participate in immediately updating data (Goodchild and Glennon, 2010).

Crowdsourced geographic information in Indonesia

The initial activities implementing crowdsourced geographic information in Indonesia carried out in June 2011 included:

1 Conducting trial and error experiments using the OSM platform to be adopted in rural and urban areas in Indonesia;
2 Identifying the need for tools and materials for the subsequent mapping of spatial objects transferred to the OSM platform; and
3 Ensuring the OSM application can be useful for the assessment of contingency planning and emergency situations.

Source: AIFDR (2014: 1)

Based on field observations for this research, the development of crowdsourced geographic information activities in Indonesia is growing rapidly, especially geo-tagging activities like travel, restaurants, locating traffic congestion, hiking and cycling route tracking. Crowdsourced geographic information activities for mapping building objects, such as editing, digitising and updating spatial data, for a whole Indonesia area through the OSM platform and the open-source GIS platforms by Volunteered Geographic Information communities in Indonesia showed increased features numbers for the period 2011–2020.

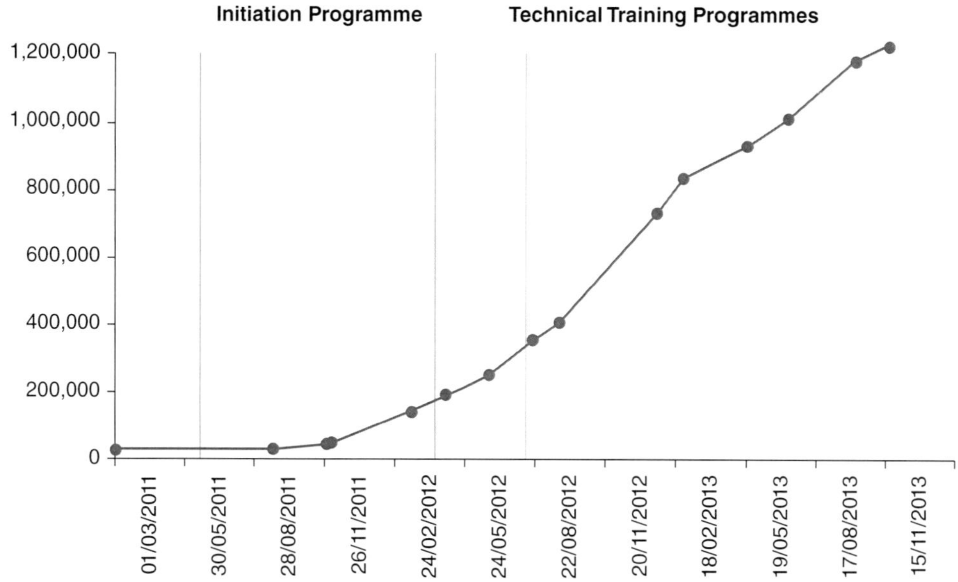

Figure 4.3 Time series of the number of buildings mapped in Indonesia through the OpenStreetMap platform in the period 2011–2013

Source: HOT Indonesia, 2020. © OpenStreetMap contributors

According to the HOT Indonesia Humanitarian OpenStreetMap Team project report in 2012 (HOT is the official name of the OpenStreetMap organisation), in the OSM pilot study of implementation between June 2011 to March 2012, 163,912 buildings were mapped. HOT Indonesia's 2013 Annual Report records that 1.3 million buildings had been mapped. In other words, nearly eight times the number of buildings were mapped in the period 2012–2013, than at the beginning of OSM operations in 2011 (Figure 4.3). By 2020, the number of buildings mapped by 2 January 2020 was 27,842,084 (Figure 4.4). For the seven years from 2013–2020 during which crowdsourced geographic information participants have contributed through the OSM platform, 21 times the number of buildings were mapped.

Strategies have been undertaken by HOT Indonesia for training OSM communities, including basic and advanced training in building geographic databases in OSM; mapathons (face-to-face groups for building spatial object mapping); and OSM mapping and consultations through social media such as Facebook and Twitter. In an effort to increase the digitisation of geographic objects mapped, HOT Indonesia held a mapping competition with rewards for members who mapped the greatest number of spatial objects.

In addition, other efforts at OSM community training were undertaken by HOT Indonesia. These included promotions to recruit new members by cooperating with government officials, schools and universities, as well as civil society organisations such as the Scouts (Pramuka) and the Indonesian Red Cross (Palang Merah Indonesia, PMI).

Building geographic databases and training and recruiting new members of the OSM community continue to be effective. According to interview respondents, voluntary digital

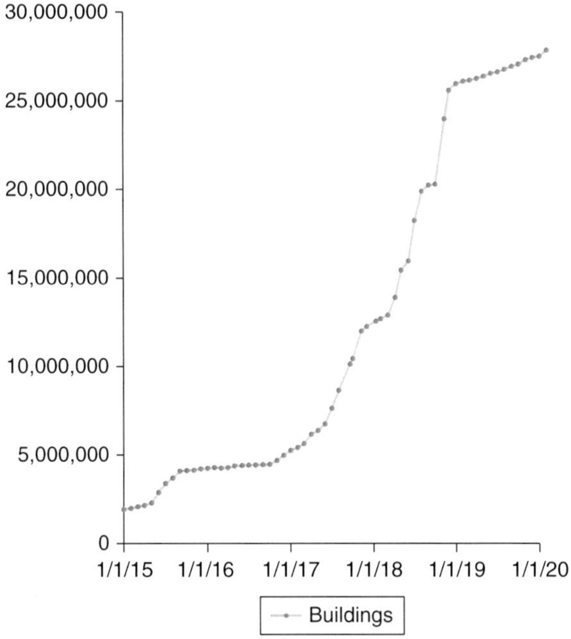

Figure 4.4 Time series of the number of buildings mapped in Indonesia through the OpenStreetMap platform in the period 2015–2020

Source: HOT Indonesia, 2020. © OpenStreetMap contributors

mapping in Indonesia is still largely project-based even though it is the communities' own initiatives to identify the objects to be mapped in their areas. When the project was over, participatory digital mapping progress was slow or stopped, because there was no supporting budget for participatory mapping community empowerment. Additionally, a lower or non-existing supporting budget will affect the awareness of the local community to keep producing and updating geographic data, except for particular persons or communities in particular areas who have a particular interest in producing and updating spatial data and information. This is an existing deficiency in building awareness of digital mapping of their own territory for communities.

Exploring a success story of crowdsourced geographic information experience to collect geographic information in Indonesia

In Indonesia, collecting geographic information activities by crowdsourced geographic information communities is dominated by OSM, and is mainly aimed at digitising road networks, land use or building parcels for the purpose of disaster management activities as one part of urban planning and management consideration in Indonesia. To explore citizen participation in the crowdsourced geographic information, this section examines the success story of crowdsourced geographic information experiences in Dompu Regency in Indonesia.

Dompu Regency is a district in West Nusa Tenggara Province, Indonesia. The regency is located at the centre of Sumbawa Island. Its area is 2,321.55 square kilometres and its population in 2020 is 236,665 or its population density is about 102 (person/km^2). The topographical condition of Dompu Regency generally can be described as partly flat (18.48%) to hilly area with land slope of 15–40% (31.55%); and 49.97% of the total area has slopes above 40% (see Figure 4.5 for Dompu Regency orientation). Land use in Dompu Regency is dominated by dry and cultivated land. Geomorphologically, there are rocky hills and dry lands where food crops cannot be grown. In general, 64% of the land in Dompu Regency is dry tropical forest, which covers a series of hills; 23% of the land is dry field; and only 11% is recorded as rice fields. Soil types in Dompu Regency include the brown Mediterranean lithosal complex, the renzina and lithosal complex with an area of 63,460 ha (Maguantara, 2006). Climatically, Dompu Regency includes areas with tropical climates with an average rainy season from October to April every year, having climate types D, E and F (according to Schmidt-Fergusson's climate classification). In the dry season, the temperature is relatively low (20 °C–30 °C) during the day and below 20 °C at night. The rainfall in the Dompu Regency area ranges from 900–1,400 mm per year with the number of rainy days ranging from 60–120 rainy days per year (Maguantara, 2006).

Figure 4.5 Map of Dompu orientation

Source: OpenStreetMap, 2020. © OpenStreetMap contributors

Crowdsourced geographic information activities created a notably unique and successful experience in Dompu Regency, according to an interview with the senior management staff of HOT Indonesia:

> There was an interesting story about crowdsourced geographic information communities in Sumbawa from 2011 to 2013, who were very enthusiastic to map out their territory. There was one village in Dompu Regency which mapped out all their buildings by their village community. Seeing how interactive community mapping results can produce good spatial data quality; the official local government, through BAPPEDA Dompu [the Planning Board of Dompu Regency] provided funds to the local community up to IDR 1 billion to map all villages under Dompu Regency using OSM.
>
> *Source*: HOT Indonesia senior management member, interview, 6 February 2015

Based on interviews with one HOT Indonesian organiser, it appears that Dompu Regency citizens are enthusiastic about participating in digitising geographic data activity on the OSM platform because the initiation project for building the geographic information database of buildings, roads and land parcels was attended by community leaders or public figures in their community.

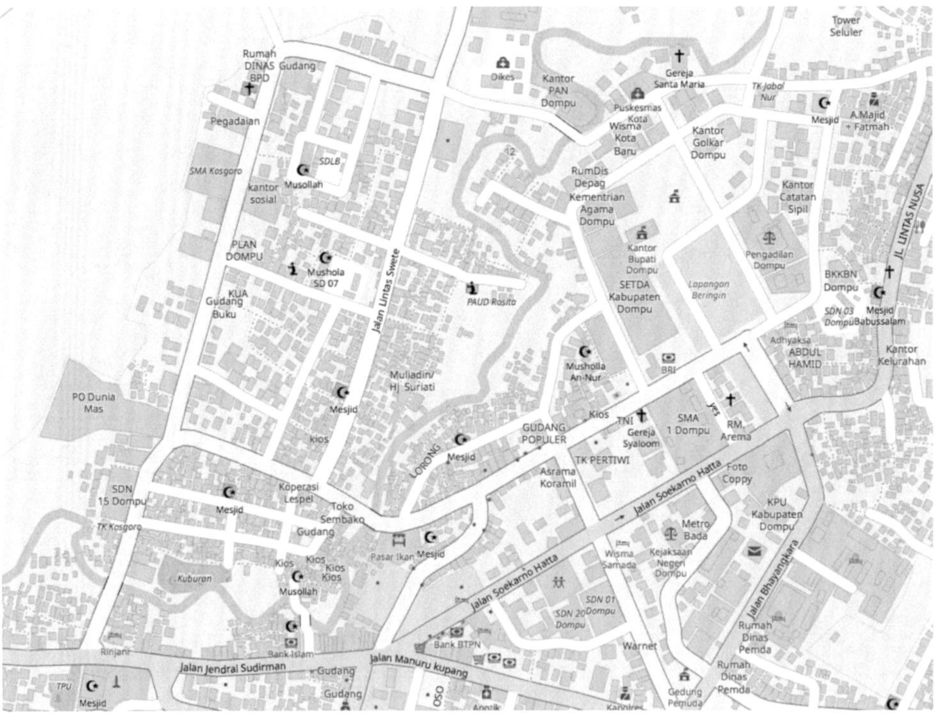

Figure 4.6 Dompu Regency – OSM platform

Source: OpenStreetMap, 2020. © OpenStreetMap contributors

The impact of the creating geographic information database was not only useful for disaster management but also for supporting socio-economic activities in the local community by having owners identify their own property, such as business and houses as well as public spaces. Thus, the community leaders who attended the initiation meeting with OSM organisers took immediate action by involving their community members in the geographic information database of buildings. The village community leaders instructed their community members to learn something new and useful so that community members' awareness was raised in the local community to voluntarily build a geographic information database (see Figure 4.6).

The phenomenon of the successful participation of the community through creating geographic information features in Dompu Regency can be explained by the 'community of practice' theory (Wenger, 1998). Wenger argues that the community which understands the importance of learning external factors – in this research context, building a geographic information database – will create awareness by itself to practise what they have learned to improve their community in daily social life.

The phenomenon can also be explained through the study on crowdsourced geographic information behaviour by Coleman et al. (2009). A crowdsourced geographic information activity could achieve success in building a geographic information database in a particular community when each person has at least one of the seven aspects of crowdsourced geographic information voluntary behaviour previously mentioned: altruism; a personal or professional interest; intellectual stimulation; social reward; enhanced personal reputation; pride of place; independent self-expression; and personal investment.

Figure 4.7 Gili Trawangan, Gili Meno and Gili Ayer building geographic information database by the OSM community

Source: OpenStreetMap, 2020. © OpenStreetMap contributors

The results of interactive mapping conducted by the crowdsourced geographic information community in Dompu were of excellent quality and provided benefits to support the Dompu Regency Government's agenda, which permitted the sizeable budget allocation discussed in the Hot Indonesia interview recorded on 6 February 2015. (In 2013, £1 was approximately equivalent to IDR 20,000, which meant the budget allocation for IDR 1 billion was valued at £50,000.)

The success story of community engagement in developing interactive mapping activity in Dompu Regency, in the Province of West Nusa Tenggara, is one example of the best qualities of crowdsourced geographic information. Furthermore, this success story had a positive impact on building a geographic information database by another crowdsourced geographic information community in the same province to map all assets within tourism areas on the Gili Trawangan, Gili Meno and Gili Ayer Islands (see Figure 4.7). They mapped all property assets in the islands to use as key information for tourism planning and environmental management.

Community empowerment through crowdsourced geographic information activities to support urban planning and management processes

Within the limited resources available, communities need decision-making based on understanding the importance of sustainability and the conflicts of interest that exist between the quality of life of a community (neighbourhood); social, economic, and environmental assets; and the potential benefits for different stakeholders in the decisions taken. In principle, there must be less bureaucratic distance between government and citizens at the level of participatory decision-making, but this also increases the intensity of communications amongst stakeholders, including researchers, experts and policymakers.

Academics can act as facilitators, helping to identify local community needs as well as mapping local social issues through collecting data and analysing the information on a crowdsourced geographic information platform. At a later stage, their knowledge of local communities, the kinds of additional data needed and their ability to access data is a crucial contribution to a community.

In terms of citizen participation, the following kinds of geographic data and information can be collected to support urban planning and management process:

1 Community inventory and asset evaluation;
2 Social resources;
3 Infrastructure resources;
4 Environmental resources; and
5 Culture and history.

Currently, GIS applications are used extensively by citizens, individually or at local community level for involvement in urban planning and management processes (these activities include collecting data, mapping, analysing and decision-making). Participation in digitalised geographic data can be categorised in four degrees or intensities (McCall, 2004): information sharing; consultation; involvement; and action.

1 **Information sharing:** One- or two-way communication between 'outside' and local community involvement.
2 **Consultation:** Consultation is more like mentoring, directed to the key problems and the kinds of training needed by the community.
3 **Involvement in decision-making by all actors:** Interaction between internal and external actors to identify priorities, analysing conditions, selecting alternatives and tools.
4 **Initiating action:** Independent initiatives driven by local communities which mobilise themselves to support relevant activities.

(McCall, 2004: 5)

This does not imply that participation must always be at maximum intensity, but the intensity should be appropriate to the task, competence and specific relationships between the actors participating in urban planning and management process.

The role of crowdsourced geographic information in urban planning and management process has been examined by McCall (2004). He argues that a local crowdsourced geographic information community does have a direct impact on the policies made by the government, so spatial information visualisation products created by a crowdsourced geographic information community are related to information sharing.

Conclusion

Crowdsourced geographic information practices in Indonesia are predominantly organised by recognised international crowdsourced geographic information management organisations, in this case OpenStreetMap (OSM). OSM activities are performed in annual projects and initially digitise geographic data on buildings, roads, bridges and other infrastructure objects in urban and rural areas to support government activities in disaster management. Assessment activities which were initiated by HOT Indonesia aimed to build a geographic information database for disaster management purposes which has created effective and efficient central government performance.

Based on the exploration of crowdsourced geographic information practices in Indonesia for this research, some key findings can be determined.

Firstly, in the OSM collaborative model, spatial data and information were validated by an OSM community itself and concerns about data and information quality could be raised by OSM members. Subsequently, an acceptance technique was created, despite the fact that the standard is not formalised as in conventional official spatial data or information created by the Indonesian National Mapping Agency (*Badan Informasi Geospasial*, BIG). Addressing the digital mapping problem, building and maintaining geodatabase-based community participation can be understood as 'communities of practice' (Wenger, 1998). The community is aware that external factors can benefit from the community and is able to establish the identity of the region. In other words, public awareness for updating spatial data voluntarily through digital mapping occurs automatically when a community is able to show how mapping geographic objects increases learning and contributes to forming community identity.

Secondly, geographic data or information sharing in the crowdsourced geographic information context in Indonesia normally disseminates geographic data at large cartographic scales (1:1,000; 1:500). In addition, metadata is automatically recorded when a crowdsourced geographic information contributor registers on the OSM platform for the first time. However, ultimately, the systematic upgrading of geographic data or information will depend on the procedures for implementing digital mapping projects.

Thirdly, geographic information database development was undertaken voluntarily by the Dompu Regency community which had been given attention by the government of Dompu Regency. The results of the interactive mapping conducted by the crowdsourced geographic information community produced excellent quality data and provided benefits to support the Dompu Regency Government's agenda.

All in all, nowadays, crowdsourcing geographic information is becoming a trend in line with the current digital information era. The fast acquisition of information in the shape of massive numbers and various formats plays a role in creating essential analysis tools for urban planning and management by providing local spatial knowledge. In the future, with the entering of the right algorithm, crowdsourcing geographic information can implement Geographic Artificial Intelligence (GeoAI) to interpret spatial patterns, so that urban management problems can be solved precisely, accurately and quickly.

References

Antoniou, V. and Skopeliti, A. (2015) 'Measures and indicators of VGI quality: An overview', *ISPRS Annals of the Photogrammetry, Remote Sensing and Spatial Information Sciences*, Vol. II-3/W5, ISPRS Geospatial Week 2015, 28 September – 3 October 2015, La Grande Motte, France.

Australia-Indonesia Facility for Disaster (AIFDR) (2014) *Expansion of InaSAFE and OpenStreetMap in Indonesia* (January 2014 – February 2015 Report). Jakarta, Indonesia.

Basiri, A., Jackson, M., Amirian, P., Pourabdollah, A., Sester, M., Winstanley, A., Moore, T. and Zhang, L. (2016) 'Quality assessment of OpenStreetMap data using trajectory mining', *Geo-spatial Information Science*, 19 (1), 56–68.

Coleman, D.J., Georgiadou, Y. and Labonte, J. (2009) 'Volunteered geographic information: The nature and motivation of producers', *International Journal of Spatial Data Infrastructures Research*, 4: 332–58.

Craig, W.J. and Elwood, S.A. (1998) 'How and why community groups use maps and geographic information', *Cartography and Geographic Information Science*, 25 (2): 95–104.

DiBiase, D., MacEachren, A.M., Krygier, J.B. and Reeves, C. (1992) 'Animation and the role of map design in scientific visualization', *Cartography and Geographic Information Science*, 19 (4): 201–14.

Elwood, S. (2006) 'Beyond cooptation or resistance: Urban spatial politics, community organizations, and GIS-based spatial narratives', *Annals of the Association of American Geographers*, 96 (2): 323–41.

Feick, R. and Roche, S. (2013) 'Understanding the value of VGI', in D. Sui, M. Goodchild and S. Elwood (eds), *Crowdsourcing Geographic Knowledge: Volunteered Geographic Information (VGI) in Theory and Practice*. London: Springer. pp. 15–30.

Fonte, C.C., Antoniou, V., Bastin, L., Estima, J., Arsanjani, J.J., Bayas, J. C. L., See, L. and Vatseva, R. (2017) 'Assessing VGI data quality', in G. Foody, L. See, S. Fritz, P. Mooney, A-M. Olteanu-Raimond, C.C. Fonte and V. Antoniou, (eds.), *Mapping and the Citizen Sensor*. London: Ubquity press. pp. 137–163.

Goodchild, M.F. (2007) 'Citizens as sensors: The world of volunteered geography', in *VGI Specialist Meeting Position Papers*. 2007: Santa Barbara, CA. http://www.ncgia.ucsb.edu/projects/vgi/docs/position/Goodchild_VGI2007.pdf (accessed December 24, 2020).

Goodchild, M.F. and Glennon, J.A. (2010) 'Crowdsourcing geographic information for disaster response: A research frontier', *International Journal of Digital Earth*, 3 (3): 231–41.

Gould, P. (1999) *Becoming a Geographer*. Syracuse, NY: Syracuse University Press.

Haklay, M. (2010) 'How good is volunteered geographical information? A comparative study of OpenStreetMap and Ordnance Survey datasets', *Journal of Environment and Planning B: Planning and Design*, (37): 682–703.

Humanitarian OpenStreetMap Team (HOT) (2012) *Community Mapping for Exposure in Indonesia* (Project Report, March 2012). https://openstreetmap.id/docs/Community_Mapping_for_Exposure_in_Indonesia_EN.pdf (accessed December 24, 2020).

Johnson, P.A. and Sieber, R.E. (2013) 'Situating the adoption of VGI by government', in D. Sui, M. Goodchild and S. Elwood (eds), *Crowdsourcing Geographic Knowledge: Volunteered Geographic Information (VGI) in Theory and Practice*. London: Springer. pp. 65–82.

Longhorn, R. and Blakemore, M. (2008) *Geographic Information: Value, Pricing, Production and Consumption*. Boca Raton, FL: CRC Press.

Maguantara, Y.N. (2006) Kerawanan Pangan dan Persoalan Yang Melingkupi: Kasus Kabupaten Dompu, Nusa Tenggara Barat (Food Insecurity and Issues Covering: The Case of Dompu Regency, West Nusa Tenggara), Jurnal Analisis Sosial, 11 (1).

McCall, M.K. (2004) 'Can participatory-GIS strengthen local-level spatial planning? Suggestions for better practice', paper presented *GISDECO 2004* on 10–12 May 2004, Johor, Malaysia. Available from https://www.researchgate.net/publication/228758277_Can_participatory-GIS_strengthen_local-level_spatial_planning_Suggestions_for_better_practice (accessed December 24, 2020).

Obermeyer, N. (2006) 'Measuring the benefits and costs of GIS', in P. Longely, M. Goodchild, D. Maguire and D. Rhind (eds), *Geographical Information Systems: Principles, Techniques, Management and Applications* (2nd edition). Hoboken, NJ: Wiley. pp. 601–10.

Roche, S. and Caron, C. (2009) 'Introduction of GIS organizaton', in S. Roche, C. Caron and M. Goodchild (eds), *Organizational Facets of GIS*. New York: Wiley.

Rouse, L.J., Bergeron, S.J. and Harris, T.M. (2007) 'Participating in the geospatial web: Collaborative mapping, social networks and participatory GIS', in A. Scharl and K. Tochtermann (eds), *The Geospatial Web: How Geo-Browsers, Social Software and the Web 2.0 are Shaping the Network Society*. London: Springer. pp. 153–8.

Schmidt, F.H and Ferguson J.H.A. (1951) *Rainfall Types Based on Wet and Dry Periods Ratio for Indonesian with Western New Guinea*. Djakarta: Kementerian Perhubungan Jawatan Meteorologi dan Geofisika.

Senaratne, H., Mobasheri, A., Loai Ali, A., Capineri, C. and Haklay, M. (2017) 'A review of volunteered geographic information quality assessment methods', *International Journal of Geographical Information Science*, 31 (1): 139–167.

Smith, R.W. (1973) 'A theoretical basis for participatory planning', *Policy Sciences*, 4 (3): 275–95.

Sui, D., Goodchild, M. and Elwood, S. (eds) (2013) *Crowdsourcing Geographic Knowledge: Volunteered Geographic Information (VGI) in Theory and Practice*. London: Springer.

Tulloch, D.L. (2007) 'Many many maps: Empowerment and online participatory mapping', *First Monday*, 12 (2). http://firstmonday.org/ojs/index.php/fm/article/view/1620/1535 (accessed December 24, 2020).

UGM and the Indonesian Humanitarian OpenStreetMap Team (2012) *Evaluation of OpenStreetMap Data in Indonesia (Case study: Yogyakarta, Surabaya, Jakarta, Bandung, Padang, and Dompu) Final Report*. Yogyakarta, Indonesia: UGM.

Wenger, E. (1998) *Communities of Practice: Learning, Meaning, and Identity*. Cambridge: Cambridge University Press.

5

DESIGN AND NARRATIVE TECHNIQUES FOR URBAN DATA VISUALISATION

Duncan A. Smith

Introduction

The impressive pace of recent urban digital technology innovation looks set to continue, with widespread implications for how people live and work in cities. Several key trends are occurring simultaneously, including the increasing availability of open data at city, national and international scales; the commercial development of spatio-temporal 'big data' and real-time data, including tracking, sensor and Internet of Things data streams; the continued improvement of analytical and visualisation software; and the increasing ease of sharing tools online and through mobile apps (Kitchin, 2014). The breadth of these changes has been labelled as a data revolution with commentators theorising how core urban services such as transport and healthcare are changing profoundly in the age of smart cities (Albino et al., 2015; Townsend, 2013).

Urban planning has continuously evolved in response to technological and political change (Batty, 2013; Hall, 2014). The promise of the developing field of urban analytics is to provide timely usable data analysis and visualisation to support urban research and planning, and potentially to engage citizens more widely in planning decision-making. While there are many emerging opportunities for innovation in planning, we can also identify some significant challenges as well. Digital divides are increasingly prominent, in terms of differences in demography, wealth and education that affect how different user groups engage with digital services (Servon, 2008). There are also institutional divides, particularly where urban tech companies control urban service data streams, creating tensions with public sector agencies which require widespread access.

This chapter focusses on using data visualisation and analytics tools in public-facing applications for non-expert audiences. Design and usability considerations are particularly important in these applications. While online tools using open data have the potential to reach large public audiences, careful design decisions are required to translate that potential into a reality. Design techniques including using interface metaphors, controlling levels of user interaction and creating data narratives are discussed. These techniques are illustrated using a series of urban data visualisation examples, classified according to levels of user interaction into exploratory, analytics and narrative tools. Task-focussed analytics tools are particularly relevant for urban planning contexts. Carefully designed maps and visualisations, responding to user questions with meaningful indicators, have the potential to reach large public audiences, with recent examples from transport and real-estate presented. Data journalism examples show how a clear narrative can help make data-driven articles and visualisations more accessible and engaging for general audiences. Structuring data narratives, editorial techniques and selecting more engaging visualisation methods can aid understanding and expand audiences for data analysis tools (Cairo, 2012; Kirk, 2016). Future urban analytics applications are likely to add more advanced narrative functionality, and to improve the ease of creating analytics and visualisation applications using cloud-based platforms.

Developing urban analytics tools

This section discusses approaches for developing urban analytics tools from a user perspective, including user interface structures, controlling the degree of interactivity provided to the user, and presents a basic typology of urban data visualisation tools.

Design techniques

A user-led approach to urban analytics means prioritising usability in digital tool development. In light of this aim, we can discuss some foundational design principles for promoting usability. A good starting point is to plan the overall interface structure and consider adopting an interface metaphor. This establishes a consistent and often familiar basis for the user experience. For example, a very common data design metaphor is 'overview-to-detail', where we begin with a big picture overview followed by 'detail on demand', where user interactions control transitions to more detailed analysis (Shneiderman, 1996). Interactive maps are ideally suited to this approach, beginning with a large scale overview map where the user can then zoom in to areas of interest (Smith, 2016). Overview-to-detail designs are also applicable for statistical analysis, beginning with aggregate statistical overview, where the user can then alter the focus or explore additional variables.

There are many alternative interface metaphors to the common overview-to-detail approach. As real-time and near real-time datasets are increasingly available, dashboard-style interfaces are increasingly prevalent, where the focus is on highlighting current trends and temporal patterns compared to archived data (Kitchin et al., 2015). Some applications focus at a small scale at the individual or neighbourhood level. This can work particularly well for rich spatio-temporal data in terms of a 'day-in-the-life' analysis, or for presenting detailed neighbourhood profiles. Another alternative interface structure is the 'small multiples' approach, where the interface consists of an array of many small visualisations for the user to overview (Tufte et al., 1990). Small multiples can be used to represent parameter spaces of many simulation results (Wilson and Dearden, 2011).

In combination with the interface structure, the degree of interactivity provided to the user is a primary design consideration, determining the dimensions of user control. A greater number of interactivity options adds to the power and flexibility of data visualisation tools. This added complexity does however increase the demands on new users learning the interface, as discussed further in the next section. A successful interface needs to anticipate and prioritise the functionality most relevant to the user, while making appropriate design decisions for any additional functionality that is less of a user priority. It is also increasingly possible to offer adaptive interfaces that can respond flexibly to the needs of users with different experience levels (Kang et al., 2003).

Another useful principle that is fundamental to data journalism is to consider how the analysis can be made more meaningful and immediate to wider audiences. This can be as simple as expressing results in data units that are more straightforward for users to understand in cases where standard scientific units are more abstract. Thinking about user relatability may also influence indicator selection, and the choice of visualisation methods used. Narrative approaches are also relevant for making data analysis more engaging and memorable, as discussed in the next section.

Levels of user interaction: exploratory, narrative and analytics approaches

In recent years, an array of different types of data visualisation tools have emerged, from dashboards to data catalogues, infographics to mapping platforms. A useful means of classifying these tools is in terms of the degree of user interaction provided, as illustrated in Table 5.1. We can consider tools based on their balance between exploratory and presentational functionality (MacEachren, 1994).

Exploratory tools are generally geared towards more complex engagement with data, such as enabling multi-dimensional analysis; allowing more sophisticated user queries; and ultimately enabling user-driven testing of ideas and hypotheses. Such functionality is complementary to the multi-dimensional and complex nature of cities, and potentially can stimulate more in-depth user engagement and understanding. The downside of this exploratory approach is that greater knowledge and interest are required by users, and there is greater risk of excluding non-expert audiences.

A contrasting approach is to follow a presentational and narrative-driven route, as used in data journalism outputs (Segel and Heer, 2010) such as articles, visualisations and spatial data narratives such as story maps (a map-based adaptation of the popular scrolling-narrative web designs). These outputs are linear in nature and guide users through a topic with a clear beginning, middle and end. Typically, a limited number of variables are discussed, and clear conclusions are presented by the author. This editorial approach is easier for general audiences to understand, and consequently can reach large numbers of users, as illustrated by very popular data journalism examples in major publications such as the *New York Times* and *the Guardian* (Cairo, 2012). It is still possible to include a degree of depth in presentational applications by employing longer narratives. Yet the potential for the user to explore their own ideas and test hypotheses is effectively excluded using this approach.

Given the different advantages of the exploratory and presentational approaches, there is a good case for developing hybrid tools that include elements of both. Analytics tools generally fit between the exploratory and presentational approaches, offering a task-focussed approach intended to quickly provide relevant analysis for decision-making (Keim et al., 2010). This narrower focus allows developers to streamline the interface to the most relevant data indicators and visualisations, thus reducing the learning curve for users and expanding the potential audience.

Another useful hybrid technique is to add narrative and commentary elements to exploratory and analytics tools. This can take the form of an introductory narrative,

Table 5.1 Data visualisation tools typology

Data visualisation tool category	Examples	Advantages	Disadvantages
Exploratory	Exploratory mapping & statistics platforms Data catalogues GIS and statistical software	Interactive Multi-variate User-led hypothesis testing	Harder to understand Specialised audience Longer development time
Narrative	Articles, Static presentations, Videos Interactive articles, Tours, Story maps	Easier to understand Wider audience Strong narrative to guide users Clear author conclusions	Linear Few variables Can struggle with complex topics
Analytics	Analytics mapping and statistics tools Dashboards	Task focussed Answer user queries on specific topic Narrower selection of relevant indicators	Requires careful design and indicator selection Can be limiting for experienced users

guiding users through the interface and highlighting the major points of interest. Narrative techniques can include interactive tours, voice-overs and movies. These approaches have some similarities to the creation of scenarios for simulation and modelling, where a series of scenarios are developed to explore future predictions, illustrating possible paths and key relationships.

Given that data visualisation tools are overwhelmingly online, there is also the potential for users to share and author their own narratives, contributing to discussion and debate. At the simplest level, users can share links and commentary on a particular chart or map including their chosen indicators and focus, typically using social media. Developers can facilitate this kind of sharing by including the user-selected visualisation parameters in the URL. The next stage is for user narratives to develop into more advanced sharing or tours and scenarios, as discussed in the next section.

Urban data visualisation examples

The following examples illustrate the urban data visualisation design approaches discussed in the previous section, classified into exploratory, analytics and narrative applications. These examples are all live websites at the time of writing, using open data to present engaging visualisation and analysis on different dimensions of urban geography and planning.

Exploratory data platforms

Exploratory tools are designed to allow users to test their own ideas and hypotheses with a broad choice of variables and functionality. This approach is powerful and flexible, but also creates considerable design challenges due to the high number of interface options presented to the user. The first example is a statistical *Cities Data Tool* developed by the Centre for Cities, an urban policy think-tank in the UK. The major strength of this tool lies in the breadth of data variables that have been brought together for the tool, with a wide range of key economic and planning indicators synthesised at the city level as shown in Figure 5.1. Important trends regarding jobs growth and regional inequalities can be analysed using user-selected charts and tables, with an impressive database of current indicators available. This tool succeeds in enabling basic user-driven hypothesis

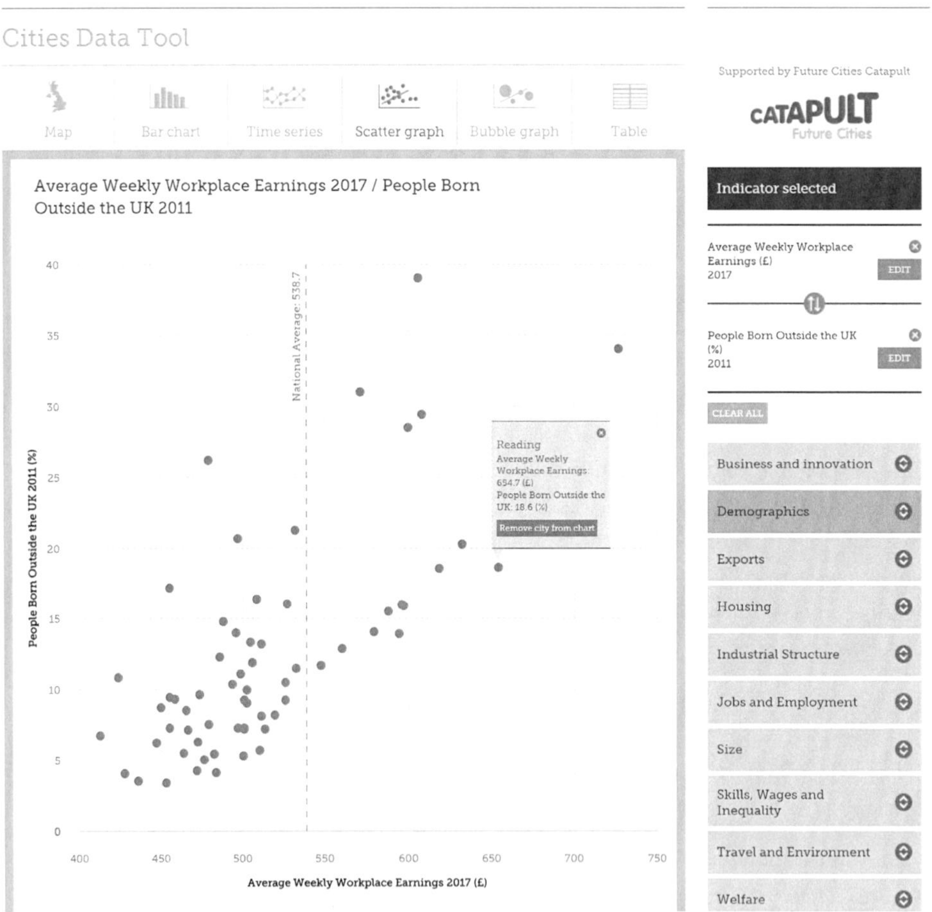

Figure 5.1 Centre for Cities City Data Tool (http://www.centreforcities.org/data-tool/ – accessed November 2019)

testing with a good degree of flexibility. The tool would benefit from some introductory narrative features to explain the interface and fundamental urban economic relationships to more novice users, as the number of options and variables could be confusing. More expert users on the other hand are likely to want more inferential statistical functionality to quantify the correlations between the variables plotted.

Carefully designed data visualisations can greatly improve usability, and this is particularly true for mapping applications. Interactive cartography allows users to explore data at a range of spatial scales. This is relevant for urban planning, where the user can move from regional trends to local neighbourhood and street level patterns. A current example of this scale-flexibility can be seen at the mapping platform developed by the Consumer Data Research Centre. This platform integrates a wide range of urban geography indicators relating to population, housing, retail, travel and health (Longley et al., 2018). The mapping data is clipped to building outlines to highlight spatial variation in the context of urban form and transport links (O'Brien and Cheshire, 2015), alongside some basic descriptive statistics. The end result is powerful and intuitive, with spatial relationships quickly highlighted at both local and city-region scales. An example is shown in Figure 5.2 where neighbourhoods with the most efficient and least efficient building energy performance are quickly highlighted. The site also functions as a data catalogue where the map indicators can be downloaded by users. The exploratory mapping approach is highly effective at highlighting spatial variation in urban socio-economic and built-environment data. The only downsides of this approach are difficulties in adding statistical analysis, and the long development time required to achieve the highly customised interface and cartography.

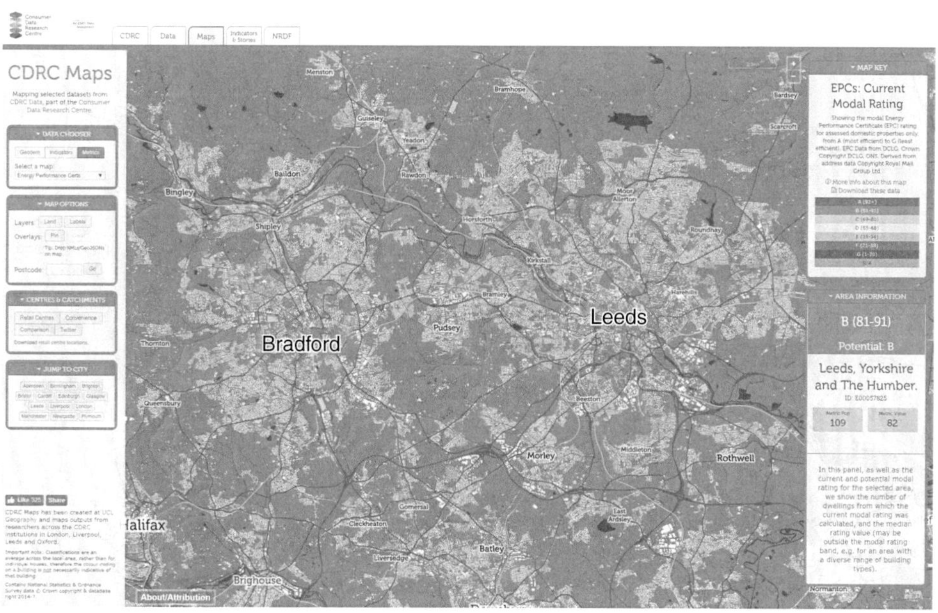

Figure 5.2 CDRC Maps: Energy performance (https://maps.cdrc.ac.uk/ – accessed November 2019)

Urban analytics examples

Exploratory platforms include a wide range of datasets, but typically are limited in terms of statistical analysis and providing quick answers on specific topics. Analytics tools generally focus on a narrower selection of indicators for a particular topic, adding relevant analytical functionality and responsiveness to user queries. Urban analytics development has been most prominent in the transport and real-estate sectors, where open data availability is well established. For example, housing affordability is an acute challenge in the UK, and the *Guardian* newspaper developed a UK housing affordability tool based on the user's salary, dynamically mapping housing affordability and calculating core statistics (see https://www.theguardian.com/society/ng-interactive/2015/sep/02/unaffordable-country-where-can-you-afford-to-buy-a-house). This tool successfully develops a meaningful affordability indicator, summarising affordability as an adjustable salary multiple, with the UK's lack of affordable housing clearly communicated. Histograms are also provided to highlight spatial variation within zones, matching the map colour scheme to aid legibility. The main shortcoming with this site relates to data currency, with a lack of recent updates meaning the analysis is missing the most recent trends.

Similar methods of user-selected queries can be used to add temporal and ranking data to mapping applications. An example of this approach is the World City Populations map 1950–2035 as shown in Figure 5.3. This map provides a quick summary of population data for large cities across the globe, including future demographic predictions, with national and international ranking statistics and global queries added for context (Smith, 2017). The visualisation uses data developed by the United Nations World Urban Prospects. International urban analytics applications are a major potential growth area, given global city competition and shared social and environmental challenges. At present, however,

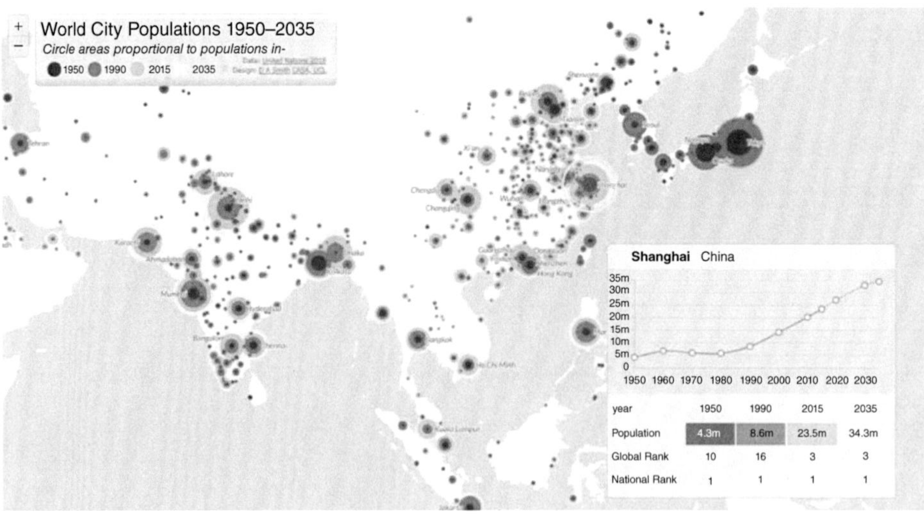

Figure 5.3 World City Populations 1950–2035 (http://luminocity3d.org/, accessed November 2019)

there are only a limited number of consistent global economic and environmental datasets at city level for multi-dimensional analysis. This will likely become less of a problem in future years as more globally integrated datasets are released (see for example efforts to integrate global urban air quality data at the World Air Quality Index project https://aqicn.org/city/london/).

More advanced analytics applications allow users to interact with models and simulations in the browser. This approach remains technically more challenging, but is increasingly feasible with advances in web technologies. Transportation is the most mature open data sector, with increasingly widely available timetable, topographic and congestion data used to power popular journey planning applications. These datasets are also being applied in urban land use planning and research. Accessibility modelling is a good example, where station and timetable data is used to analyse location specific travel opportunities and likely future travel patterns (Antrim and Barbeau, 2013). The Transport for London Time Mapping tool shown in Figure 5.4 enables interactive accessibility

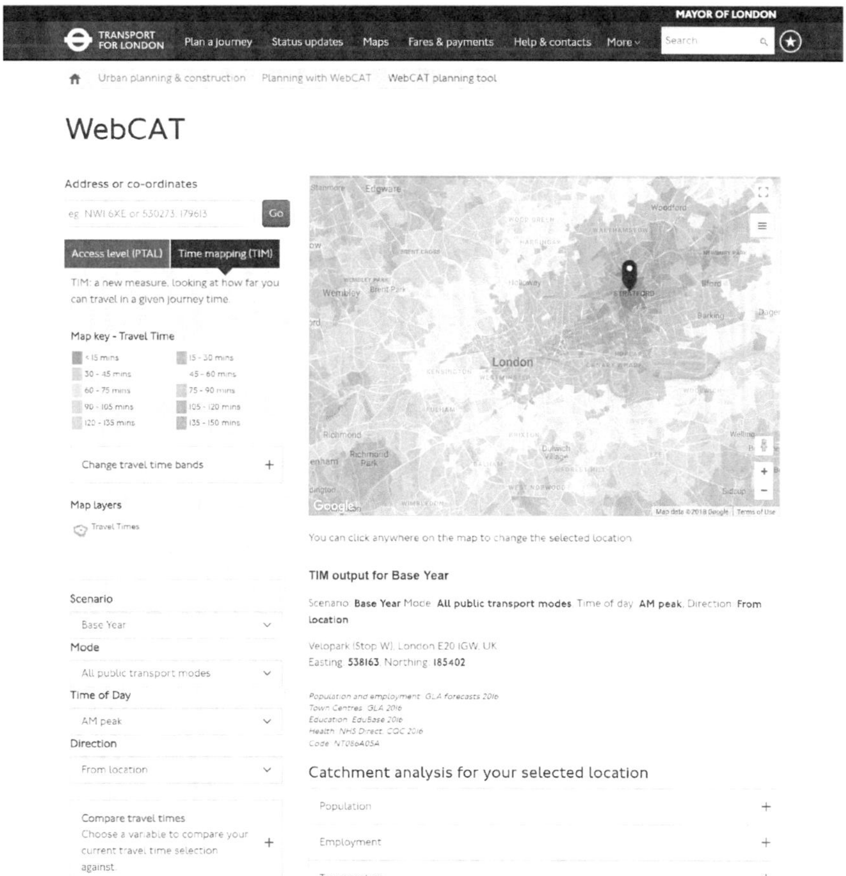

Figure 5.4 Transport for London Time Mapping Tool (https://tfl.gov.uk/WebCat, accessed November 2019) © Transport for London

mapping and catchment analysis by different travel modes, based on detailed network modelling (Inayathusein and Cooper, 2018). This type of modelling feeds into the London planning process for prioritising development sites at areas of high public transport accessibility. The Transport for London tool also includes predicted future accessibility for scenario years based on planned infrastructure development. At present the full dynamic possibilities of accessibility modelling are not comprehensively realised, but the potential of interactive online modelling is clearly shown by this tool.

Spatial data narrative examples

We have seen in the previous examples how urban data visualisation is expanding, with exploratory and analytical applications being developed in various sectors. Thus far the tools discussed have not featured significant narrative functionality. Adding introductory tours and overviews can be very effective in improving usability and engaging non-expert audiences. The main growth of narrative data analysis has been in data journalism, with many mainstream news and financial publications adding data visualisations to online articles (Cairo, 2012). The data journalism approach can also be seen in urban analysis contexts. Metro Monitor by the Brookings Institute,[1] is a good example of combining text and interactive infographics, analysing urban economic trends in the US and internationally. Another interesting development integrating GIS with data journalism functionality is Environment Systems Research Institute (ESRI)'s Story Maps platform.[2] This application integrates interactive web maps with linear scrolling presentations and animations based on a series of templates.

These examples take the form of linear articles and presentations. It is interesting to look at more hybrid examples of adding narrative functionality to fully interactive data visualisations. A more comprehensive integration can be seen in the work of multimedia companies such as Kiln,[3] where voice-over stories and tours are added to interactive data visualisations. This approach has been used in several very popular data visualisations such as the Carbon Map, and the Ship Map shown in Figure 5.5. The Ship Map visualises a very large international shipping database, revealing spatial and temporal patterns to inform energy and sustainability analysis. The introductory

[1]https://www.brookings.edu/research/metro-monitor-2018/ (accessed December 21, 2020).

[2]https://storymaps.arcgis.com/en/ (accessed December 21, 2020).

[3]https://www.kiln.digital/ (accessed December 21, 2020).

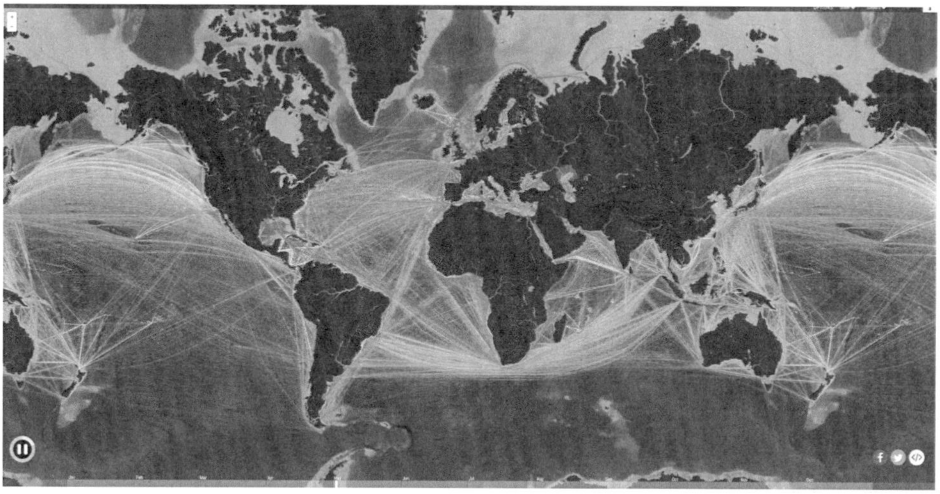

Figure 5.5 2012 Ship Map by Kiln and UCL Energy Institute (https://www.shipmap.org/ – accessed November 2019)

narrative explains the aims of the project, and then presents highlights, zooming to areas of interest and switching indicators, all intended to stimulate users to explore the visualisation in more detail. What is particularly successful in this interface design is that the introductory narrative route is presented while maintaining user control of the interface, so that the user can easily switch between these narrative and exploratory roles.

The use of data narratives is ideally suited to explaining changes through time, where past, present and planned future states can be discussed. A strong example of this spatial data narrative approach can be seen in the work of CarbonBrief,[4] who provide sustainability related journalism and analysis for energy industries. Their recent data visualisation work successfully combines interactive mapping with clear narratives describing trends and future states based on current planning policy. The example shown in Figure 5.6 visualises changes in the electricity supply mix for the UK over the last decade, showing rising levels of renewables being installed and planned for 2030. Data transitions through time are controlled through vertical scrolling with the energy supply map, descriptive text and animated chart of the UK energy mix all changing in response to the user navigating through past and future states. This provides an intuitive and engaging method of explaining an important and relatively complicated planning topic that is undergoing significant change.

[4]https://www.carbonbrief.org/ (accessed December 21, 2020).

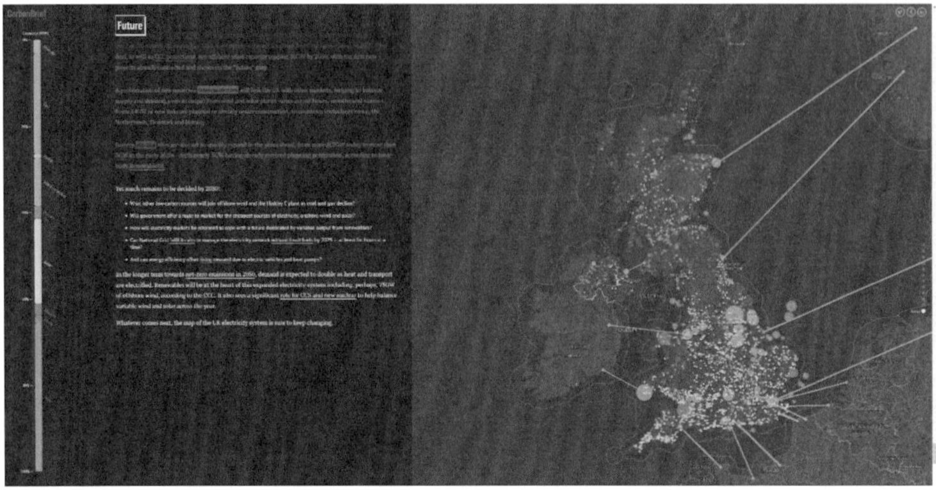

Figure 5.6 UK electricity supply transformation by CarbonBrief (https://interactive.carbonbrief.org/how-uk-transformed-electricity-supply-decade/ – accessed November 2019)

Future tools and applications

The improving functionality of core web technologies such as HTML and JavaScript (increasingly server-side as well as client-side) is enabling more powerful web applications to be developed. These advances are in combination with improved graphics standards and libraries, such as the WebGL standard increasingly popular in interactive mapping and data visualisation, which allows more ambitious graphics and animations to be rendered in standard web browsers and apps. Together these advances translate into greater potential for web visualisation and analytics.

In addition to more technical advances, an important trend is the continuing development of cloud services for data visualisation, which provide tools and templates to build data visualisations with greatly reduced requirements for programming skills (Smith, 2016). This can greatly expand the number of users able to create data visualisations and tools, including urban planners who are not specialised in IT. Sophisticated mapping services such as Mapbox and Carto fall into this category, as do popular data analysis services such as Tableau and DataWrapper, while ESRI is also expanding in this area. The adaptable template approach used by many of these services enables common types of data visualisation to be created and customised quickly by many users. Cloud services can also be used to add greater narrative functionality more easily. One of the more advanced platforms in this regard is Flourish,[5] designed as a data visualisation cloud

[5]https://flourish.studio/ (accessed December 21, 2020).

service for journalists, which allows users to easily build presentation functionality into any of their visualisation templates using a keyframe approach, similar to movie editing software. This approach means that data narrative functionality is built into the platform and can be easily developed and customised by users.

The increasing ease with which data visualisations can be created also offers greater potential for multiple interfaces for the same tool, for different levels of user experience. The aim of adaptive interfaces is to create multiple visualisations for different users. This flexible approach could potentially address problems of visualisation legibility for some public groups, and the need to ensure the widest possible audience for public-facing tools. It also offers an approach that does not require as much design compromise for expert audiences.

Conclusion

Urban data availability continues to expand from public sector open data sources and from spatio-temporal big data streams. There are therefore greater opportunities to apply these datasets to support urban analysis and planning tasks, particularly as advances in web technology and cloud services are lowering the technical hurdles for achieving high quality visualisation and analytics functionality.

This chapter focussed on public facing applications, looking to provide urban data visualisation for general public audiences. Design and usability considerations are particularly important where public engagement is a priority and the aim is to reach large audiences. Several useful design techniques were discussed including adopting interface metaphors, carefully controlling the level of user interaction and functionality, and using narrative techniques to guide users through interfaces. These techniques were illustrated using a series of urban data visualisation examples, classified accordingly by user interaction levels into exploratory, analytics and narrative tools.

Exploratory tools offer users the greatest flexibility and scope for user-driven hypothesis testing. This approach requires a broad range of data and sophisticated interface design to be successful, and is generally more applicable for specialised audiences. Analytics tools are task focussed, with functionality reduced to a particular user question, which can help simplify interfaces for general audiences. There have been significant advances in analytics tools for urban sectors such as transport and real estate, with clear relevance for urban planning contexts. Examples were provided showing interactive mapping and statistics being used in housing affordability analysis and accessibility analysis tools. Carefully designed cartography and visualisations, with meaningful indicators responding to user queries were illustrated.

Narrative techniques are generally under-utilised in urban data visualisation. Data journalism examples show how a clear data story can make articles and visualisations more accessible for general audiences, and explain major relationships and trends. It is possible to add narrative introductions to exploratory and analytics tools so that data visualisations can meet expectations for both general and more advanced users. Future applications are likely to add more advanced storytelling and narrative functionality, particularly in scenario and modelling applications where future trends need to be explained, and to expand possibilities for user-developed narratives.

References

Albino, V., Berardi, U. and Dangelico', R.M. (2015) 'Smart cities: Definitions, dimensions, performance, and initiatives', *Journal of Urban Technology*, 22 (1): 3–21. doi.org/10.1080/10630732.2014.942092

Antrim, A. and Barbeau, S.J. (2013) 'The many uses of GTFS data – opening the door to transit and multimodal applications', Location-Aware Information Systems Laboratory at the University of South Florida. http://www.locationaware.usf.edu/wp-content/uploads/2010/02/The-Many-Uses-of-GTFS-Data-–-ITS-America-submission-abbreviated.pdf (accessed December 27, 2020).

Batty, M. (2013) 'Big data, smart cities and city planning', *Dialogues in Human Geography*, 3: 274–9.

Cairo, A. (2012) *The Functional Art: An Introduction to Information Graphics and Visualization*. Berkeley, CA: New Riders.

Hall, P. (2014) *Cities of Tomorrow: An Intellectual History of Urban Planning and Design Since 1880*. Hoboken, NJ: John Wiley & Sons.

Inayathusein, A. and Cooper, S. (2018) *London's Accessibility Indicators: Strengths, Weaknesses, Challenges* (Discussion Paper). Paris: OECD International Transport Forum. Available from https://www.itf-oecd.org/london-accessibility-indicators (accessed December 27, 2020).

Kang, H., Plaisant, C. and Shneiderman, B. (2003) 'New approaches to help users get started with visual interfaces: Multi-layered interfaces and integrated initial guidance', in *Proceedings of the 2003 Annual National Conference on Digital Government Research*. Boston, MA: Digital Government Society of North America. pp. 1–6.

Keim, D.A., Kohlhammer, J., Ellis, G. and Mansmann, F. (2010) *Mastering the Information Age – Solving Problems with Visual Analytics*. Goslar: Eurographics Association. Available from https://diglib.eg.org/handle/10.2312/14803 (accessed December 27, 2020).

Kirk, A. (2016) *Data Visualisation: A Handbook for Data Driven Design*. London: Sage.

Kitchin, R. (2014) *The Data Revolution: Big Data, Open Data, Data Infrastructures and Their Consequences*. London: SAGE.

Kitchin, R., Lauriault, T.P. and McArdle, G. (2015) 'Knowing and governing cities through urban indicators, city benchmarking and real-time dashboards', *Regional Studies, Regional Science*, 2 (1): 6–28. doi.org/10.1080/21681376.2014.983149

Longley, P., Cheshire, J. and Singleton, A. (2018) *Consumer Data Research*. London: UCL Press. Available from https://library.oapen.org/handle/20.500.12657/29973 (accessed December 27, 2020).

MacEachren, A.M. (1994) 'Visualization in modern cartography: Setting the agenda', in A.M. MacEachren and D.R.F. Taylor (eds), *Visualization in Modern Cartography*. Kidlington: Elsevier Science.

O'Brien, O. and Cheshire, J. (2015) 'Interactive mapping for large, open demographic datasets using familiar geographical features', *Journal of Maps*, 12 (4): 676–83. doi.org/10.1080/17445647.2015.1060183

Segel, E. and Heer, J. (2010) 'Narrative visualization: Telling stories with data', *IEEE Transactions on Visualization and Computer Graphics*, 16: 1139–48. doi.org/10.1109/TVCG.2010.179

Servon, L.J. (2008) *Bridging the Digital Divide: Technology, Community and Public Policy*. Hoboken, NJ: John Wiley & Sons.
Shneiderman, B. (1996) 'The eyes have it: A task by data type taxonomy for information visualizations', in *Proceedings 1996 IEEE Symposium on Visual Languages*. pp. 336–43. https://ieeexplore.ieee.org/document/545307 (accessed December 27, 2020).
Smith, D.A. (2016) 'Online interactive thematic mapping: Applications and techniques for socio-economic research', *Computers, Environment and Urban Systems*, 57: 106–17. doi.org/10.1016/j.compenvurbsys.2016.01.002
Smith, D.A. (2017) 'World city populations 1950–2030: Proportional circle time series map', *Environment and Planning A: Economy and Space*, 49 (1): 3–5. doi.org/10.1177/0308518X16641414
Townsend, A.M. (2013) *Smart Cities: Big Data, Civic Hackers, and the Quest for a New Utopia*. London: WW Norton.
Tufte, E.R., Goeler, N.H. and Benson, R. (1990) *Envisioning Information*. Cheshire, CT: Graphics Press.
Wilson, A. and Dearden, J. (2011) 'Phase transitions and path dependence in urban evolution', *Journal of Geographical Systems*, 13: 1–16.

6

CITY DASHBOARDS AND 3D GEOSPATIAL TECHNOLOGIES FOR URBAN PLANNING AND MANAGEMENT

Oliver Dawkins, Rob Kitchin, Gareth W. Young and Tomasz Zawadzki

Introduction

Data have long been generated about cities to make sense of them, manage services and infrastructure, solve urban problems, and produce urban policy and plans. Indeed, managing data across the data life cycle (generating, handling, storing, analysing, archiving, sharing and destroying) is a key activity of city administrations and other public agencies. Traditionally, data are generated as a by-product of core administrative functions (such as routine bureaucracy), as a core function of infrastructure or service (such as traffic management), or through routine and commissioned surveys (such as household surveys or a census). These data provide key inputs for strategic development planning (such as creating local area plans), assessing proposed planning applications, and operational governance (such as the delivery of services).

Since the early 1990s, and through the adoption of Agenda 21 of the United Nations Conference on Environment and Development (UNCED) in particular, cities have been mandated to systematically monitor sustainable development and adopt evidence-based decision-making. To that end, cities have sought to gather and use available data to compile city indicators (recurrent quantified measures) that enable key aspects of urban life (economy, labour markets, health, education, transport, housing, etc.) to be tracked over time (Innes and Booher, 2000; Maclaren, 1996; Van Assche et al., 2010). New managerialist approaches to the public sector emerged which promoted the use of indicators to monitor the performance of cities and city administrations and track the implementation and consequences of policy initiatives in order to make public services more efficient, effective, transparent, and value for money (Behn, 2014; Kitchin et al., 2015). Urban indicators also enable benchmarking that establishes how well a neighbourhood/city is performing vis-a-vis other locales or against best practices (Huggins, 2009). More recently, the open government agenda, and the drive to create open government data, has sought to make both the underlying data and urban indicators open for the public to scrutinise and use (Kitchin, 2014a). Also, such initiatives often seek to make operational, real-time data (such as environmental sensing, weather sensors, bike share, car park spaces, real-time passenger information) available through various web services and APIs.

These core datasets and indicators provide key inputs for applied urban data analysis intended for both professional audiences (planners, managers, technocrats, policymakers) and for public consumption. One of the key means by which the results of these analyses are disseminated is through city dashboards. Here, a suite of visual devices ranging from numeric displays to analytical charts is used to display indicator data through a common graphic interface. This may also incorporate interactive elements, enabling users to query specific data. More recently, city dashboards have been complemented by the development of 3D geospatial tools that transpose city-data onto three-dimensional landscape representations of the city. In this chapter, we discuss the form of applied data analysis conducted with city dashboards and emerging 3D geospatial tools. We contend that these spatial media provide a valuable means to make sense of often complex data, especially for those that lack the skills to create their own data visualisations, though they are not without critique and shortcomings. We discuss some of those criticisms and indicate ways in which they might be addressed.

City dashboards for urban planning and management

City dashboards provide a visual means to organise and interact with data. They use visual analytics – dynamic and interactive graphics (e.g. gauges, traffic lights, meters, arrows, bar charts, graphs, maps; see Figure 6.1) – to display and communicate information about the performance, structure, pattern, and trends of cities. Key data are displayed on a screen and, in many cases, can be interacted with (e.g. selecting, filtering and querying data; zooming in/out, panning and overlaying; changing the type of visualisation, or simultaneously visualising the same datasets in different ways to explore their dimensions and identify their relationships). Because the data used are recurrent, quantitative measures, many of the visualisations that are used show change over time and are updated as new data are released. By utilising the power of the visual to summarise and convey a large amount of information, dashboards enable a user to quickly and effectively explore the characteristics and structure of datasets to identify patterns and interpret trends. As such, they act as cognitive tools that improve a user's 'span of control' over voluminous, varied, and quickly transitioning data (Brath and Peters, 2004). Indeed, city dashboards are becoming increasingly popular with city governments and agencies because they collate different streams of data and diverse sets of indicators into one system, with a complement of additional tools to visualise, query, and analyse them. Moreover, they allow users to track and compare changes over time and space, and in the case of real-time data, the here-and-now of different phenomena. The power and utility of city dashboards are premised on their claims to show the state of play of cities through objective, trustworthy, factual data; to translate the messiness and complexities of cities into rational, detailed, systematic, ordered forms of knowledge (Kitchin et al., 2015).

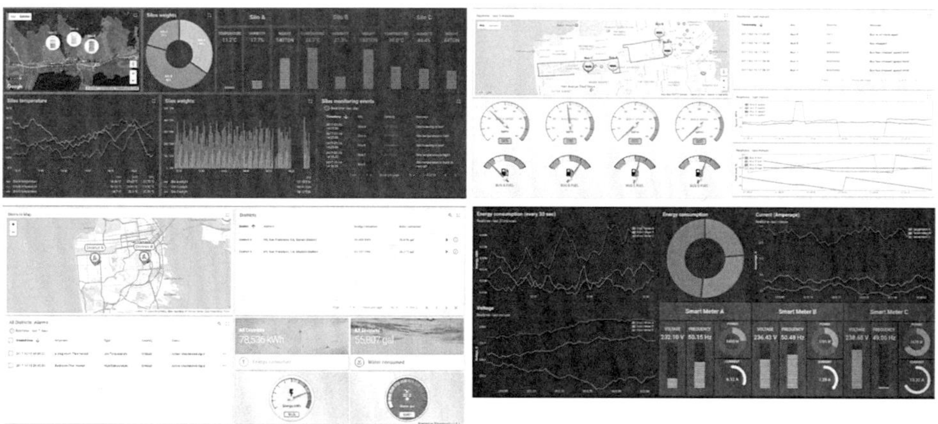

Figure 6.1 Examples of types of interactive graphics used in city dashboards

Source: Thingsboard, 'Real-Time IoT Dashboards' examples; https://thingsboard.io/ (accessed December 28, 2020).

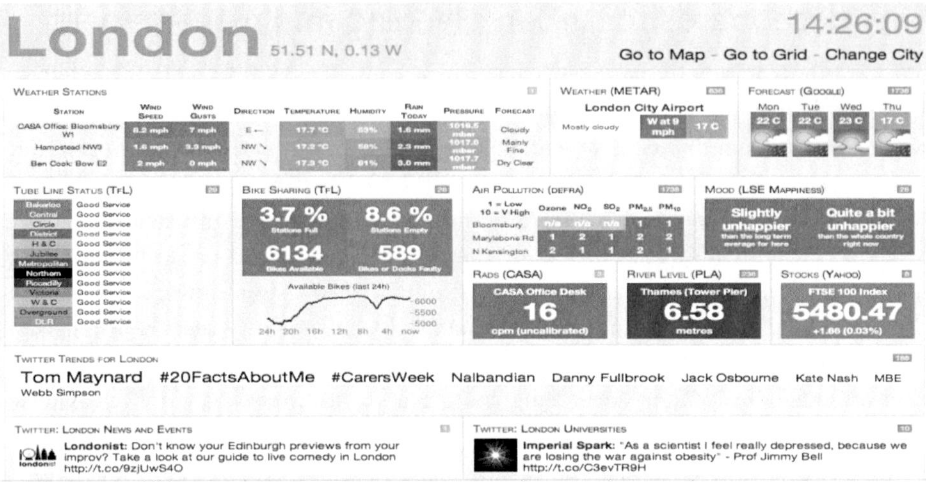

Figure 6.2 City-at-a-glance dashboard

Source: CASA, London Dashboard, http://citydashboard.org/london.

In some cases, the dashboard acts like that of a car or plane, with relevant data 'consolidated and arranged on a single screen so the information can be monitored at a glance' (Few, 2006: 34). An example of such a dashboard is the London dashboard created by CASA at University College London (see Figure 6.2). Analytical dashboards are more extensive in scope and are hierarchically organised to enable a plethora of interrelated dashboards to be navigated (typically on a thematic basis, such as planning or health), and afford summary-to-detail exploration or 'drill down' within a single system (Rivard and Cogswell, 2004) (see Figure 6.3). Analytical dashboards are more likely to include map interfaces that enable the spatial distribution of data to be examined, with these maps typically being interactive. Both types of these dashboards are common in urban control rooms, but they are also increasingly being displayed in mayor's offices, public buildings, and made accessible to the general public via dedicated websites.

While city dashboards have much utility, they also have several shortcomings and criticisms. Kitchin and McArdle (2017) detail six main concerns, which they frame with a set of questions:

- Epistemology: How are insight and value derived from city dashboards?
- Scope and access: How comprehensive and open are city dashboards?
- Veracity and validity: To what extent can we trust city dashboards?
- Usability and literacy: How comprehensible and useable are city dashboards?
- Use and utility: What are applications and value of city dashboards?
- Ethics: How can we ensure that dashboards do not cause harm and are used in a socially responsible manner?

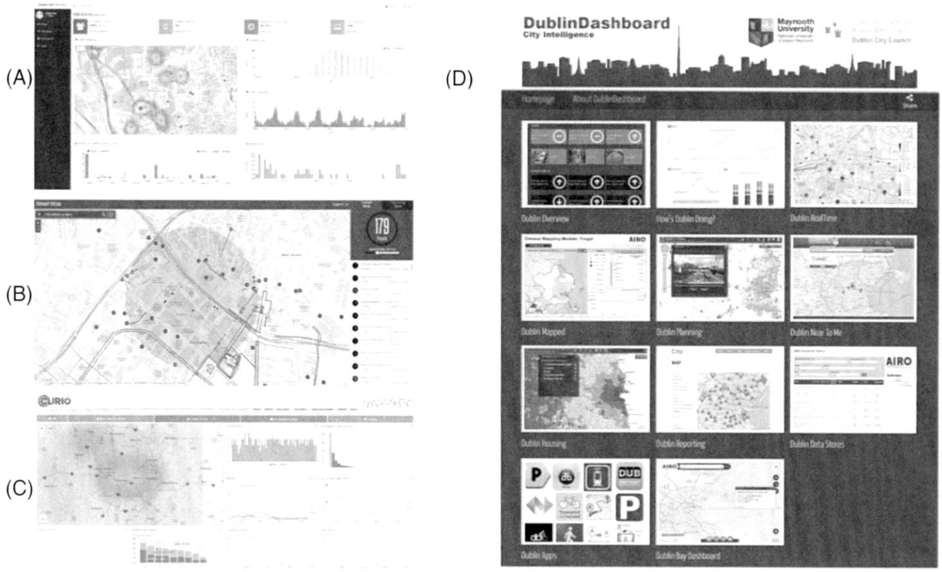

Figure 6.3 Analytical city dashboards

Source: A. Madrid dashboard (ceiboard.dit.upm.es/dashboard/); B. Los Angeles dashboard (https://navigatela.lacity.org/streetwize/); C. Columbus dashboard (http://curio.osu.edu); D. Dublin dashboard (www.dublindashboard.ie).

Their analysis builds on a long history of critique concerning the politics and praxes of the development and workings of spatial media such as GIS, location-based services and locative social media (Kitchin et al., 2017; Pickles, 1994; Sheppard, 2005), raising several fundamental and instrumental concerns about how city dashboards produce knowledge about cities and how they are used in urban planning and management. Rather than reject the use of city dashboards outright, Kitchin and McArdle (2017) instead prefer to recognise their utility and value as a mode of communication and means of making sense of the city. Their caveat is that for city dashboards to reach their full potential, issues relating to the access, quality, comprehensibility and limited scope of data will need to be addressed, as well as a more fundamental reframing of their implied epistemology through the adoption of ideas from critical GIS and radical statistics.

Building a city dashboard with user engagement

The Building City Dashboards project[1] has sought to extend initial work on the Dublin and Cork dashboards[2] by addressing the critiques identified in their initial development.

[1]dashboards.maynoothuniversity.ie (accessed December 21, 2020).

[2]dublindashboard.ie and corkdashboard.ie (accessed December 28, 2020).

Specifically, this has involved a fundamental rethink in the presentation of visual analytics and the tools provided to users, primarily based on user engagement research and requirements analysis. Typically, city dashboards are engineered as data portals that perform specific, pre-set functions with little thought given to the holistic effects of their functionality, usability, and user experience. This has become apparent in our user interviews where respondents were invited to provide feedback on four different city dashboards (Young and Kitchin, 2020).[3] It was found that many participants did not have sufficient data literacy, familiarity and competence with the use of dashboards to find relevant materials and interpret the data visualisations presented to them. The first principle of Hansen's 'User Engineering Principles' is to 'KNOW THE USER' (Hansen, 1971). In the context of city dashboards this means that their creators should make efforts to consider the specific requirements, skills, and knowledge of their intended users. As fundamental as this principle is, feedback from our interviews indicates that knowledge and understanding of user requirements is assumed without meaningful engagement of potential users in the process of scoping and design. Speedy delivery of data dashboards can have great utility in addressing short-term exigencies, but in the context of city dashboards this can be tokenistic and compromise their ongoing relevance for professional stakeholders and citizens. Creating sustainable and open city dashboards requires an informed awareness of how and why different groups of users engage with data through city dashboards to learn, think about, and frame questions about typical urban concerns. For example, our research demonstrated that complex, proprietary GIS software was too difficult to use and the functionalities provided were too specialised for many users and use cases. While a skilled user with strong data literacy and familiarity with the tools could gain significant insight from the dashboards and visualisations produced by these tools, most other users struggled to extract value from the experience, abandoned their sessions, and are unlikely to return. More generally, while city dashboards profess to bring data and insight to the masses, in reality, they are used by a small cohort of advanced users, in the main because this is the only group with the sufficient skills to use the tools and interpret their outputs. However, they were not the only group with an interest in evidence-based approaches to understanding the city.

A key issue is to produce dashboards that can be used by novice users but also still provide the tools required for more specialised uses. While acknowledging the diversity of users, pragmatic steps must be taken to develop actionable designs that address differing user requirements and cognitive styles, enabling designers to effectively determine the appropriateness of presenting tabular information over graphs, words in place of numbers, or structured rather than open-ended systems (Shneiderman, 2010). Our research identified two complementary approaches. The first is an engagement with the graphic, map and web design literature with reference to established guidelines for design and navigation

[3]Dublin, New York, London and Hawaii's dashboards.

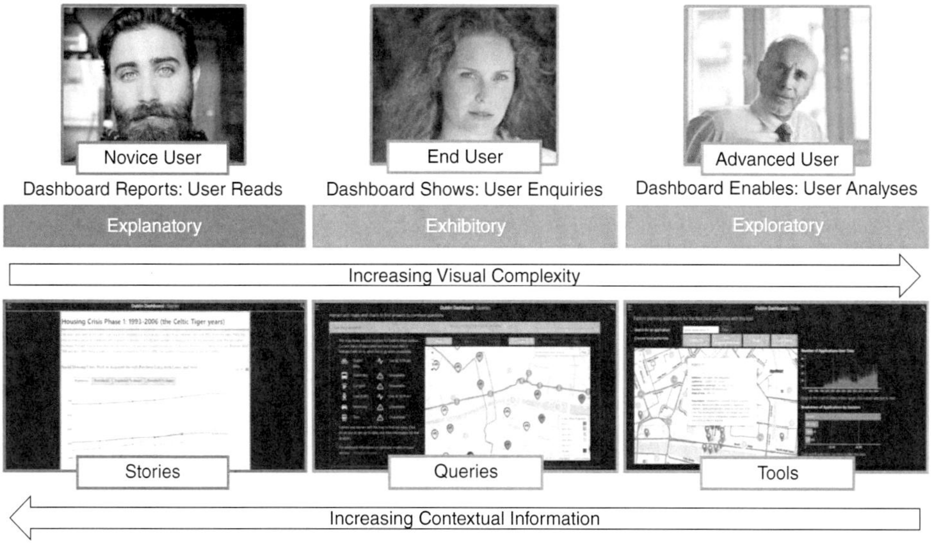

Figure 6.4 Personas

(e.g. Bollini, 2017; Kelleher and Wagener, 2011; Shneiderman, 2010; Tufte, 2001). The second involved the creation of three personas (Josh, Jane, and Geoff), each representing a potential type of user ('novice', 'end-user' and 'power user') defined along two axes – experience (casual to professional) and domain knowledge (simple to complex) (see Figure 6.4). An 'end-user' is typically someone such as a planner or policymaker who has a reasonable level of data literacy but is not a professional analyst with advanced skills. In re-designing the Dublin dashboard, we have sought to be mindful of Josh, Jane and Geoff, and their respective profiles and to consider the balance between the complexity of analysis and contextual information; between explanation and exploration. In addition to a more standard thematic organisation of information we have sought to structure the dashboard around three further modes of presentation – stories (simple data visualisations with accompanying explanatory text), tasks (for more complex and comparative data visualisations), tools (additional data analytics functions beyond data visualisation and access to the underlying data), also depicted in Figure 6.4 (Young et al. 2021). In addition to re-designing traditional city dashboards, we have also been exploring the potential of multimodal data interfaces, including using 3D environments.

3D geospatial technologies for urban planning and management

While our daily experience of cities is structured in three dimensions, the way we have traditionally made sense of and planned them has been in 2D with the aid of maps, plans and statistical plots. Except for physical scale models, a rarity due to the labour and expense involved in their creation, our understanding of cities and urban form in 3D has been limited to the views and vantage points offered by architectural drawings or, more recently, digitally manipulated photomontages. Film and computer animation have introduced duration and movement into visual media, but the experience they conveyed was typically constrained to predetermined paths and itineraries. This began to change with the introduction of real-time rendering and interaction in 3D video games and simulations through the 1990s and early 2000s (Whyte, 2002). These limitations are now in the process of being overcome through the development of new interaction technologies such as virtual reality (VR) and augmented reality (AR), new means of organising, storing and sharing spatial data at different scales, such as building information modelling (BIM) and city information modelling (CIM), and the ongoing improvement of methods for generating 3D spatial data using LiDAR scanning and photogrammetry (Valencia et al., 2015). Each of these technologies is spatially-intelligent to the extent that the underlying data can be geo-referenced, can host other data that are geo-referenced (such as the locations of specific objects, infrastructure, and services, or more abstract administrative records that can be linked to particular locations and addresses), and can thus be queried concerning spatial relationships between elements.

These technologies and data have the potential to significantly disrupt existing administrative, planning, architectural, construction and real estate practices. They can enable users to immerse themselves in, explore, interact, query, and experiment with spatially-intelligent virtual environments. Viewed on their own or overlaid on the physical environment they afford a greater understanding of a location, better decisions and policy-making, and improving operational efficiencies (Lock et al., 2019; Young et al., 2017). They create systems similar to Davide Gelernter's 'Mirror Worlds'; computer-generated models in which 'the whole city is visible on a single screen in a dense, live, pulsing, swarming, moving, changing picture' (1991: 30). While affording users the ability to 'dive deeper' into the detail, much like our city dashboards, the overriding value of mirror worlds for Gelernter derived from the overview or 'topsight' they provide; the ability to see the 'big picture' or 'image of the whole' (1991). Through the combination of 3D modelling and real-time sensing enabled through the Internet of Things, this vision is now being realised in the promotion of 'digital twin' as a means of using 3D models for the monitoring and operational management of buildings and infrastructure in real-time (Dawkins et al., 2018). From a professional perspective, digital twin systems provide responsive means for managing resources, energy consumption and cost in the development of engineered artefacts including buildings. Concerning planning and communication, the use of iconic 3D models which represent an object in the world have long been held to be less abstract and more readily understandable than the 2D maps and charts such as those in our dashboards. At the same time their use can support an epistemically 'naive realism' (Smallman and St John, 2005): the higher the degree of detail they contain, and the more closely they imitate the physical environment through the application of realistic textures and lighting effects (such as shadows that change with the time of day), the more they convey a sense of verisimilitude and the appearance of 'truth' that may in fact be lacking (Kitchin and Dodge, 2002).

These issues are gaining new relevance with the emergence of enhanced 3D spatial media like VR (virtual reality) and AR (augmented reality) which enable novel means of representing, querying, and interacting with urban data, both within and beyond the fields of planning and urban design (Kitchin et al., 2020). Immersive VR simulates a user's presence in a 3D immersive virtual environment (IVE) using a head-mounted display (HMD) with an embedded stereoscopic screen and other equipment such as earphones and handsets facilitating an embodied interaction within the IVE. The development of VR systems can be traced back to the 1960s (Sutherland, 1968), however, until the mid-2010s use of such systems was experimental and confined to military, research, and high-end industrial contexts due to cumbersome, expensive equipment, limited data, and sub-optimal processing and graphics (Jerald, 2015). Each of these issues has been addressed to some degree, with more affordable, and portable commercial headsets such as the Oculus Rift and HTC Vive. Importantly, VR places the user directly into the IVE by phenomenologically engulfing their senses so as to bestow a sense of immersion and experience of presence when necessary technical and experiential conditions relating to tracking, latency, the field of view and opportunities for interaction are sufficiently met (Curran, 2018). In the context of cities, VR can enable users to explore a 3D simulation of the urban environment at 1:1 scale, and experience and interact with it as if they were present in that scene, at least for controllable variables like spatial layout, visual appearance, ambient noise, and other sensations afforded by peripheral devices such as treadmills or haptic feedback devices. Through networking VR also provides opportunities for such experiences to be shared without requiring the co-presence of participants. Planning and urban design present prime use cases for VR because it enables planners and architects to envision how their proposed designs will fit into the existing environs (their aesthetic properties, impact on sightlines and overshadowing), simulate their relationship to various phenomena (internal energy performance or external pedestrian movements), determine their likely consequences (impact on local demographics, transportation, and economics) and iterate their designs in response (Fisher and Unwin, 2002; Kitchin et al., 2019; Portman et al., 2015).

AR and mixed realities provide a different kind of experience by integrating digital content with the user's actual environment so that they are able to observe and interact with both simultaneously. Originally designed to help factory workers assemble complex items by overlaying blueprints and instructions (Caudell and Mizell, 1992), AR can now be used to enable collaboration by allowing several planners to view and interact with the same 3D model concurrently without compromising their ability to communicate face-to-face. It can also be used in-situ, on a construction site for example, as a means of validating the design and verifying the construction process. The Pokémon Go phenomenon of 2016 also demonstrated the viability of AR for use by citizens on their smartphones as they move through the city. By using their smartphone's GPS functionality while exploring the city they revealed location-based data in the form of collectible digital creatures, which could then be superimposed on either a stylised environment or the 'real world' as seen through the smartphone's camera. That data might instead be interesting tourist information linked to a monument, planning files associated with a particular building, or the locations of utility assets otherwise hidden beneath the city streets, indeed any data that can be spatially referenced to the landscape and represented by a text, image or 3D overlay. In this way, AR exhibits a great deal of flexibility in terms of potential contexts for use and range of applications. As a medium AR is less demanding than VR in terms of its physical requirements for interaction and its monopolisation of the user's sensorium. It also therefore has far greater potential for achieving ubiquity in public space. This in turn may have far-reaching consequences for the way we perceive our environment and interact in public space as suggested by designer Keiichi Matsuda's concept film *Hyper-Reality* (2016).

Enabling these experiences will require new data infrastructures. For the built environment a BIM is a 3D model of the physical and functional aspects of a building or facility which has additional information attached such as details of components, temporal data for construction and installation, and their overall cost (Crotty, 2011). More than a simple visualisation tool, BIM enables the full build cycle of a project to be viewed and queried within one model, reducing the need to coordinate and search hundreds of separate plans, sections and elevations, which can instead be extracted directly from the BIM. It allows users to dynamically update and recalculate scheduling and quantities of materials with changes in design or specifications, and to track supplier details and automatically generate orders for materials. Further, a BIM can be shared across multiple project partners, enabling closer collaboration through the alignment and tracking of complex workflows. After construction, BIM can then be used to coordinate on-going facility maintenance. BIM has become widespread across the urban development sector for managing large construction projects because it enhances project intelligence and management, streamlines processes and produces significant cost savings. For this reason, practitioners emphasise that, beyond its instantiation in any 3D model, BIM is a process (Azhar, 2011).

CIM seeks to extend the vision of BIM beyond buildings, facilities, and infrastructure to the city scale (Thompson et al., 2016). While it would be advantageous to link BIMs together to form a CIM, in reality very few parts of a city as yet have an associated BIM. There are also significant challenges in the integration, management, and visualisation of detailed BIM data at the urban scale. A putative standard for CIM exists in the form of the OGC open standard for 3D city models CityGML.[4] While the standard has gained a highly engaged academic audience concentrated in mainland Europe, wider adoption has been slow. As a result, relatively few CityGML models are being actively used and maintained on an operational basis. One reason has been limited interoperability with other more widely adopted 3D data formats and associated modelling and visualisation applications (Biljecki and Arroyo Ohori, 2015). Another has been the limited support and documentation available for extending the existing CityGML schema to accommodate the development of new applications (Biljecki et al., 2018). Assessments of the standard's

---

[4]https://www.ogc.org/standards/citygml

performance in Singapore's flagship project Virtual Singapore (Soon and Khoo, 2017), combined with responses to the planned release of CityGML 3.0, are likely to have a significant impact on determining its future adoption.

Instead of waiting for standards to match up with the needs of developers and urban authorities, many prototype CIMs are being pieced together from a variety of existing and more highly generalised data sources: layering infrastructure and land-use information, administrative and statistical indicator data, and real-time operational data onto the 3D model. In this way the CIM seeks to act as a 3D spatial data infrastructure to be used for city-wide asset management, examining the spatial relationships between phenomena, and running simulations to determine the behaviour of a range of urban phenomena under different conditions. In essence, CIM seeks to provide an integrated city operating system, distributing urban intelligence across places, sectors, and infrastructures, in order to enable informed decision-making as to how a city is managed in the here-and-now and planned for the future (Marvin and Luque-Ayala, 2017).

Thus far, CIMs are in the prototype phase. Several universities have developed working platforms: Vilo by the Centre for Advanced Spatial Analysis at University College London; CIVAL at Future Cities Laboratory at ETH Zürich; Digital Campus Innovation by Carleton Immersive Media Studio at Carleton University; RAISE project by the Design Lab at the Queensland University of Technology; VGN by Northumbria University; VirtualOulu by the Centre for Ubiquitous Computing at the University of Oulu. Meanwhile, companies such as Cityzenith, VuCity and Dassault Systèmes, partners in Virtual Singapore, have developed nascent commercial products. Still others like A-VR, RealSim, SmarterBetterCities, CyberCity3D, and virtualcitySYSTEMS have been exploring different use cases and applications or developing supporting data and services (see Figure 6.5). Existing CIMs typically work in a desktop environment, and only in a few cases with immersive VR/AR or on mobile devices. The 3D models they incorporate can be derived by the import of existing CAD files, extrusion from 2D building footprints, procedural generation, digitising from stereoscopic images, or conversion from LiDAR or aerial photogrammetry. The models are generally placed on top of map tiles or aerial photographs, though digital terrain models of varying detail and quality may also be used to provide topographic context.

The functionality available on each platform varies. Most adopt a view from above with interactive navigation achieved with a rotating and panning virtual camera similar to those used in 3D modelling and visualisation. Some platforms also provide the kinds of first-person views encountered in computer games with ground movement, flying or teleportation. Some platforms include simulations relating to sunlight, shadows, sightlines, airflow and water physics, people and traffic movements. Some platforms enable users to query the attributes of objects such as building and associated planning information. It may also be possible to toggle different layers of objects and information such as trees and street furniture and to import new models representing planned buildings and infrastructure. In other cases, geographic data can be imported and visualised and may include support for animated views of dynamic real-time data.

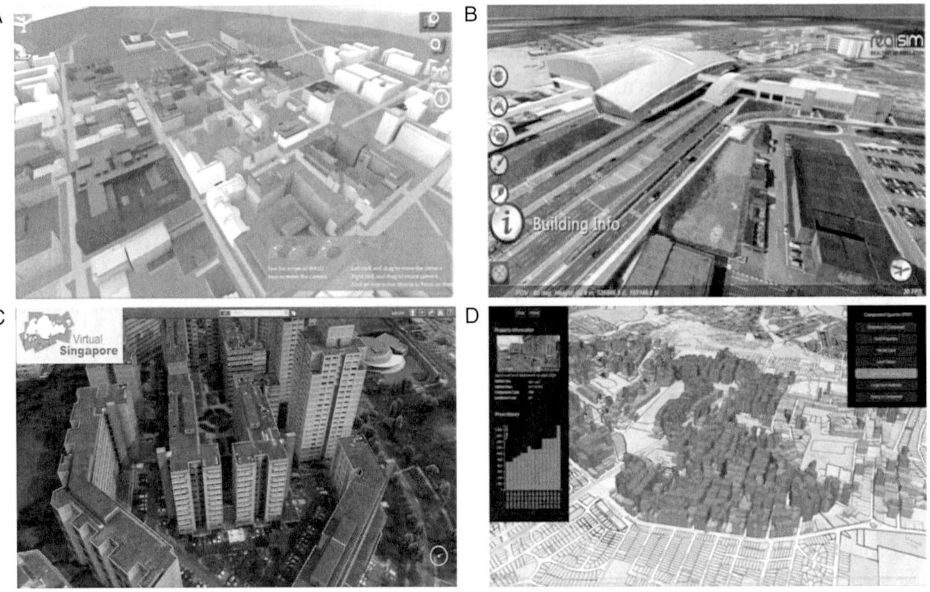

Figure 6.5 Example CIMs

A: VirtualOulu (image from https://ubicomp.oulu.fi/infrastructure-virtualoulu/); B: RealSim (image from https://realsim.ie/); C: Virtual Singapore (image from https://www.nrf.gov.sg/programmes/virtual-singapore); D: RAISE (image from www.research.qut.edu.au).

Refiguring the Dublin city dashboard in 3D

We have been developing our 3D platform for the Dublin dashboard (see Figure 6.6). Taking planning as our case study, we are interested in the extent to which 3D visual analytics might provide useful tools for planners at all stages of planning work (e.g. creating strategic, long term plans; public consultation; evaluating planning applications; enforcement). To that end, we have been building desktop, VR (Oculus Rift & HTC Vive), and AR (Microsoft HoloLens) applications. Like other such platforms, we have added tools that enable features to be queried (e.g. clicking on a building to pull up its planning file), new buildings to be added to the landscape, and visualisations of the effects of shadows and the impact of rising water levels. We are also examining ways to import a suite of georeferenced data to create a kind of 3D GIS, and to be able to explore the temporal aspects of such data to show change over time. Here, we are particularly interested in utilising the third dimension to create a new dynamic form of existing kinds of geographic data landscapes or 'datascapes' (Amoroso, 2010). Where the 2D map limits the display of layered data to a single layer at a time due to occlusion, and confines data relating to a specific place to a point or polygon, the 3D medium enables multiple layers and dimensions of data to be viewed simultaneously. In a real-time 3D environment, the represented landscape becomes the medium for the display of data, and those data in

Figure 6.6 Dublin 3D dashboard

Top left: Dublin Docklands in 3D; Top right: Dublin bike scheme real-time visualisation; Bottom left: Dublin air quality monitoring visualisation; Bottom right: Users with Microsoft Hololens mixed reality (interaction concept image) and HTC Vive virtual reality.

turn form a new synthetic landscape. Areas defined by geographic boundaries become volumes, lines describing paths become walls, and points become towers, each of which can be subdivided, or can incorporate or provide the support for newly derived data and their visualisation (e.g. points-towers that describe not only a magnitude but a standard deviation). In this sense, in VR the abstraction from territory to map becomes partially elided through the creation of virtual landscapes (Dodge and Kitchin, 2000), though AR can also operate in a Borgesian way, with the map directly draped on the territory. As such, 3D spatial media renderings open interesting questions concerning the mapping of cities that require further philosophical thought (Kitchin and Dodge, 2002).

Like city dashboards, 3D geospatial tools are not without their challenges. At a basic level, it will be difficult to implement comprehensive models in many city jurisdictions in the short-to-medium term unless there are significant improvements in data availability and quality. Moreover, the tools to produce and combine high quality, interactive media with geographic data are still lacking. While it is relatively straightforward to display a 3D model, importing other geospatial data and adding visual analytics into these models is more complex – in the main because the games engines used to facilitate real-time multimodal interaction through VR and AR are not designed to act like GISs (in fact they typically make different assumptions about the choice of coordinate reference system and the degree of precision required for calculating spatial location). Likewise, 3D geospatial applications will require critical attention as to their production and epistemology and the social, political and economic effects of their deployment. As with mapping in general and other spatial media, 3D geospatial technologies present the world as a systematic, ordered, commonsensical form of knowledge, made even more powerful because their visual form replicates how the world appears (3D), rather than being abstracted into 2D. Moreover, they enable us to project the 'present future' (Adam, 2004), which is to imagine the city as it might become by placing potential new buildings and infrastructure into the models, which then become the blueprints for creating that future. Yet, as Rose et al. (2014) have argued, 3D architectural renderings and city images are far from apolitical. They are in fact laden with politics and ideology designed to promote certain urban visions. Further, 3D geospatial technologies generally deal with location, geo-referenced factual data, and urban form, and largely ignore the metaphysical aspects of human life, and the role of politics, ideology, social structures, capital, and culture in shaping cities (Kitchin, 2014b). The danger is then, that if used in isolation, they decontextualise a city from its history, its political economy, its wider interconnections and interdependencies, the wider set of social, economic and environmental relations, and the everyday experiences of people living in the city. As CIMs gain wider adoption and begin to have tangible impacts on the planning and management of cities, citizens may regret the epistemic and ontological commitments embodied in the programming of their front-facing functionality and the schemas structuring their underlying data if they go unexamined and unchallenged.

As Dawkins (2017) argues, while many mirror-world-like platforms and applications are aimed at professional users in the fields of architecture, engineering, construction, urban planning, and asset and operations management, such systems are not inherently limited to these professional groups. As with participatory and critical GIS, and counter-mapping and citizen science initiatives (Elwood, 2006; Haklay, 2013; Peluso, 1995), 3D geospatial technologies can be used reflexively and in conjunction with other forms of knowledge and praxes to provide a more situated view and can be appropriated by individuals and communities in pursuit of their ends. He discusses ViLo, a 3D spatial environment of London created by CASA at University College London, which enables the exploration of dynamic data visualisations of various urban events as they occur in real-time (such as the movement of buses and trains, the availability of bikes at bike-share stations, sensor readings, and social media postings at particular locations). Used with AR the system, ViLo could enable the discovery of and interaction with data and other digitally mediated aspects of the city. In relation to the user's actual position in the city, spatial relationships that seem abstract on a two-dimensional map would become relatable with regard to the scale and orientation of the human body; and digital artefacts like data visualisations could be intuitively incorporated into their user's own spatial and sensory field of reference through their augmented field of view. Unlike the closed 'dataspaces' of control rooms, that system presently operates on entirely publicly accessible and open data and could form the basis of collaborative workspaces in which citizens could explore and formulate plans for their own local environments. With support, instruction, and realistic expectations, such systems can be pieced together from free-to-use and open-source software. In this way systems like ViLo point the way and contribute to the opening of a space of collaborative digital agency and participation.

Conclusion

City dashboards are becoming increasingly popular as a means of monitoring and presenting recurrent and real-time data about various aspects of cities and using the information to make sense of and manage services and infrastructures. With their use of various visual analytics, they constitute a very salient example of applied data analysis for urban planning and management. That said, they are not without critique and our research on the Building City Dashboards project is seeking to tackle the six issues identified by Kitchin and McArdle (2017) relating to epistemology, scope and access to data, data quality and ecological fallacies, usability and data literacy, appropriate application, and socially responsible design and use. In particular, there is still much work to be done to improve user experiences and to formulate general design principles for communicating indicator data, time-series data, and real-time data (Stehle and Kitchin, 2020; Young and Kitchin, 2020) that are common for other forms of visualisations such as maps. Particularly, there is a need for approaches that are guided by user feedback and requirements and which try to cater for a range of potential users (Young et al., 2021). There is also much scope for moving beyond visual analytics to include a wider set of data analytics, statistical modelling, and generate and evaluate predictions, simulations, and optimisations, which is another core aspect of our research.

It is also the case that emerging 3D geospatial technologies are providing opportunities to create new ways of visualising cities and their associated data. Their representative qualities (as 3D rather than 2D), immersive nature, and interactive spatial intelligence are starting to generate a diverse set of applications concerning planning, city operations management, asset management, architecture, construction, policy formulation, autonomous vehicles, citizen and tourist applications, and urban play and games. However, immersive VR, AR, CIM, and other technologies reliant on 3D data, such as autonomous cars, are still a few years from reaching the tipping point where their use becomes mainstream. In part, this is because the technology and its application to domains such as planning are still evolving, with affordable headsets only recently becoming available and use cases still in the experimental rather than the commercial phase. In part, this is because development platforms are yet to emerge that satisfactorily combine the real-time rendering capabilities of game engines with the GIS-like ability to store and handle geospatial data in native formats and coordinate systems with the requisite spatial precision. It is also still relatively difficult to find publicly accessible 3D city models that are sufficiently highly detailed to meet user expectations while maintaining the kinds of cross-application compatibility required by developers. That said, several national mapping agencies have started to generate new national 3D datasets derived from LiDAR data, and there are data with the variable quality available through OpenStreetMap and commercial vendors. Large corporations, such as Google and Nokia, have created or are in the process of creating global 3D datasets for a variety of projected uses, including autonomous vehicles, though these are not publicly available. Besides, there is an increasing number of project-based 3D models generated for specific purposes (e.g. asset management and urban development) by companies and city administrations. The quality, veracity, and fidelity of these data are improving markedly, especially for high-value markets such as large-scale construction.

In short, while there are still a set of fundamental and technical challenges to be overcome and critique to be addressed, city dashboards and 3D geospatial technologies have the potential to provide a set of tools for urban managers *and* citizens for making sense of, planning and managing cities, and to do so increasingly in real-time. Whether these will constitute disruptive innovations that will radically transform the praxes of domains such as planning, only time will tell.

Acknowledgements

The research for this article was provided by a Science Foundation Ireland grant, 'Building City Dashboards' (15/IA/3090). Some of the material is updated and extended from a blog post for the *International Journal of Urban and Regional Research* 'Spotlight On' series, 'Disruptive 3D geospatial technologies for planning and managing cities'. https://www.ijurr.org/spotlight-on/disruptive-urban-technologies/disruptive-3d-geospatial-technologies-for-planning-and-managing-cities/

References

Adam, B. (2004) *Time*. Cambridge: Polity Press.

Amoroso, N. (2010) *The Exposed City: Mapping the Urban Invisibles*. London: Routledge.

Azhar, S. (2011) 'Building information modeling (BIM): Trends, benefits, risks, and challenges for the AEC industry', *Leadership and Management in Engineering*, 11 (3): 241–52.

Behn, R.D. (2014) *The PerformanceStat Potential: A Leadership Strategy for Producing*. New York, NY: Brookings Institution Press/Ash Center.

Biljecki, F. and Arroyo Ohori, K. (2015) 'Automatic semantic-preserving conversion between OBJ and CityGML', in *UDMV15: Eurographics Workshop on Urban Data modelling and Visualisation*, Delft, The Netherlands, 23 November. pp. 25–30. doi.org/10.2312/udmv.20151345

Biljecki, F., Kumar, K. and Nagel, C. (2018) 'CityGML Application Domain Extension (ADE): Overview of developments', *Open Geospatial Data, Software and Standards*, 3 (1): article #13.

Bollini, B. (2017) 'Beautiful interfaces: From user experience to user interface design', *The Design Journal*, 20 (1): 89–101.

Brath, R. and Peters, M. (2004) 'Dashboard design: Why design is important', *DM Direct*, 85: article #1011285-1. Available from http://cs.furman.edu/~pbatchelor/csc105/articles/TUN_DM_ONLINE.pdf

Caudell, T.P. and Mizell, D.W. (1992) 'Augmented reality: An application of heads-up display technology to manual manufacturing processes,' in *Proceedings of the Twenty-Fifth Hawaii International Conference on System Sciences, Vol 2*. Kauai, HI, USA. pp. 659–69.

Crotty, R. (2011) *The Impact of Building Information Modelling: Transforming Construction*. London: Routledge.

Curran, N. (2018) 'Factors of immersion', in K. Norman and J. Kirakowski (eds), *The Wiley Handbook of Human Computer Interaction* (Volume 1). London: Wiley. pp. 239–54.
Dawkins, O. (2017) 'Opening urban mirror worlds: Possibilities for participation in digital urban data spaces', *Proceedings of the Whose Right to the Smart City conference*. Available from https://www.researchgate.net/publication/328497812_Opening_Urban_Mirror_Worlds_Possibilities_for_Participation_in_Digital_Urban_Dataspaces (accessed December 28, 2020).
Dawkins, O., Dennet, A. and Hudson-Smith, A. (2018) 'Living with a digital twin: Operational management and engagement using IoT and Mixed Realities at UCL's Here East Campus on the Queen Elizabeth Olympic Park', in *Proceedings of the 26th annual GIScience Research UK conference: GISRUK 2018*. GIS Research UK (GISRUK): University of Leicester, UK.
Dodge, M. and Kitchin, R. (2000) *Mapping Cyberspace*. London: Routledge.
Elwood, S. (2006) 'Critical issues in participatory GIS', *Transactions in GIS*, 10 (5): 693–708.
Few, S. (2006) *Information Dashboard Design: The Effective Visual Communication of Data*. Sebastopol: O'Reilly.
Fisher, P. and Unwin, D. (eds) (2002) *Virtual Reality in Geography*. London: Taylor and Francis.
Gelernter, D. (1991) *Mirror Worlds: Or the Day Software Puts the Universe in a Shoebox...How It Will Happen and What It Will Mean*. Oxford: Oxford University Press.
Haklay, M. (2013) 'Citizen science and volunteered geographic information', in D. Sui, S. Elwood and M. Goodchild (eds), *Crowdsourcing Geographic Knowledge*. Amsterdam: Springer. pp. 105–22.
Hansen, W.J. (1971) 'User engineering principles for interactive systems', in Fall Joint Computer Conference, Atlantic City, 1971. Available from https://www.computer.org/csdl/proceedings/afips/1971/5078/00/50780523.pdf (accessed December 28, 2020).
Huggins, R. (2009) 'Regional competitive intelligence: Benchmarking and policy-making', *Regional Studies*, 44 (5): 639–58.
Innes, J. and Booher, D.E. (2000) 'Indicators for sustainable communities: A strategy building on complexity theory and distributed intelligence', *Planning Theory & Practice*, 1 (2): 173–86.
Jerald, J. (2015) *The VR Book: Human-centered Design for Virtual Reality*. San Rafael, CA: Morgan & Claypool.
Kelleher, C. and Wagener, T. (2011) 'Ten guidelines for effective data visualization in scientific publications', *Environmental Modelling & Software*, 26 (6): 822–7.
Kitchin, R. (2014a) *The Data Revolution: Big Data, Open Data, Data Infrastructures and Their Consequences*. London: Sage.
Kitchin, R. (2014b) 'The real-time city? Big data and smart urbanism', *GeoJournal*, 79 (1): 1–14.
Kitchin, R. and Dodge, M. (2002) 'There's no there there: Virtual reality, space and geographic visualisation', in P. Fisher and D. Unwin (eds), *Virtual Reality in Geography*. London: Taylor and Francis. pp. 341–61.
Kitchin, R. and McArdle, G. (2017) 'Urban data and city dashboards: Six key issues', in R. Kitchin, G. McArdle and T. Lauriault (eds), *Data and the City*. London: Routledge. pp. 111–26.
Kitchin, R., Dawkins, D. and Young, G. (2019) 'Prospects for an intelligent planning system', *Planning Theory & Practice*, 20 (4): 575–603.
Kitchin, R., Lauriault, T.P. and McArdle, G. (2015) 'Knowing and governing cities through urban indicators, city benchmarking and real-time dashboards', *Regional Studies, Regional Science* 2: 1–28.
Kitchin, R., Lauriault, T.P. and Wilson, M. (eds) (2017) *Understanding Spatial Media*. London: Sage.

Kitchin, R., Young, G. and Dawkins, O. (2020) 'Planning and 3D spatial media: Progress, prospects, and the knowledge and experiences of local government planners', *Planning Theory and Practice*, 20(4): 595–9.

Lock, O., Bednarz, T., Leao, S.Z. and Pettit, C. (2019) 'A review and reframing of participatory urban dashboards', *City, Culture and Society*, 20: article #100294.

Maclaren, V.W. (1996) 'Urban sustainability reporting', *Journal of the American Planning Association*, 62 (2): 184–202.

Marvin, S. and Luque-Ayala, A. (2017) 'Urban operating systems: Diagramming the city', *International Journal of Urban and Regional Research*, 41 (1): 84–103.

Matsuda, K. (2016) *Hyper-Reality* [Video]. https://vimeo.com/166807261 (accessed December 28, 2020).

Peluso, N.L. (1995) 'Whose woods are these? Counter-mapping forest territories in Kalimantan, Indonesia', *Antipode*, 27 (4): 383–406.

Pickles, J. (ed.) (1994) *Ground Truth: The Social Implications of Geographic Information Systems*. New York, NY: Guildford Press.

Portman, M.E., Natapov, A. and Fisher-Gewirtzman, D. (2015) 'To go where no man has gone before: Virtual reality in architecture, landscape architecture and environmental planning', *Computers, Environment and Urban Systems* 54: 376–84.

Rivard, K. and Cogswell, D. (2004) 'Are you drowning in BI reports? Using analytical dashboards to cut through the clutter', *DM Review*, April 2004. https://web.archive.org/web/20130123045925/http://www.advizorsolutions.com/press/Cut%20Through%20The%20Clutter.pdf

Rose, G., Degen, M. and Melhuish, C. (2014) 'Networks, interfaces and computer-generated images: Learning from digital visualisations of urban redevelopment projects', *Environment and Planning D: Society and Space*, 32: 386–403.

Sheppard, E. (2005) 'Knowledge production through Critical GIS: Genealogy and prospects', *Cartographica* 40 (4): 5–21.

Shneiderman, B. (2010) *Designing the User Interface: Strategies for Effective Human-Computer Interaction*. Delhi: Pearson Education India.

Smallman, H.S. and St John, M. (2005) 'Naive realism: Misplaced faith in realistic displays', *Ergonomics in Design*, 13 (3): 6–13.

Soon, K.H. and Khoo, V.H.S. (2017) 'CityGML modelling for Singapore 3D national mapping', *The International Archives of Photogrammetry, Remote Sensing and Spatial Information Sciences*, 42: 37–42.

Stehle, S. and Kitchin, R. (2020) 'Real-time and archival data visualisation techniques in city dashboards', *International Journal of Geographic Information Science*, 34 (2): 344–66.

Sutherland, I.E. (1968) 'A head-mounted three dimensional display', *Proceedings of the December 9–11, 1968, Fall Joint Computer Conference, Part I*. ACM. pp. 757–64. doi.org/10.1145/1476589.1476686

Thompson, E.M., Greenhalgh, P., Muldoon-Smith, K., Charlton, J. and Dolník, M. (2016) 'Planners in the future city: Using city information modelling to support planners as market actors', *Urban Planning* 1 (1): 79–94.

Tufte, E.R. (2001) 'Graphical excellence,' in E.R. Tufte, *The Visual Display of Quantitative Information*. Cheshire, CT: Graphics Press. pp. 13–52.

Valencia, J., Muñoz-Nieto, A. and Rodríguez-Gonzálvez, P. (2015) 'Virtual modeling for cities of the future. State-of-the art and future challenges', *The International Archives of Photogrammetry, Remote Sensing and Spatial Information Sciences*, 40 (5): 179–85.

Van Assche, J., Block, T. and Reynaert, H. (2010) 'Can community indicators live up to their expectations? The case of the Flemish city monitor for livable and sustainable urban development', *Applied Research Quality Life*, 5: 341–52.

Whyte, J. (2002) *Virtual Reality and the Built Environment*. Oxford: Architectural Press.

Young, G. and Kitchin, R. (2020) 'Creating design guidelines for building city dashboards from a user's perspectives', *International Journal of Human-Computer Studies*, 140: article #102429.

Young, G., Kitchin, R. and Naji, J. (2021) 'Building city dashboards for different types of user', *Journal of Urban Technology*, 28 (1–2): 289–309. doi.org/10.1080/10630732.2020.1759994

Young, G.W., Naji, J., Charlton, M., Brunsdon, C. and Kitchin, R. (2017) 'Future cities and multimodalities: How multimodal technologies can improve smart-citizen engagement with city dashboards', in *ISU Talks #05: Future Cities conference*, at the Institute for Sustainable Urbanism, TU Braunschweig.

7

DATA ANALYTICS, URBAN FORM AND CLIMATE CHANGE: THE URBAN CLIMATE MAP

Chao Ren, Edward Ng, Jason Wai Po Tse, Pak Shing Yeung, Jimmy Chi Hung Fung, Gerald Mills, Jason Ching, Benjamin Bechtel and Linda See

List of abbreviations

WUDAPT: World Urban Database and Access Portal Tools

SDGs: Sustainable Development Goals

IPCC: Intergovernmental Panel on Climate Change

WRF: Weather Research and Forecasting

PRD: Pearl River Delta

GE: Google Earth

LCZ: Local Climate Zone

UHI: urban heat island

Background and challenges

We are living in an urbanised world, yet we know very little about most cities on the planet, especially how they are built (urban form) and occupied (urban function). Urban residents are especially vulnerable to current and future climate hazards because of the concentration of people, infrastructure and social-economic activities in landscapes that are exposed to sea-level rise, river flooding, storms, heatwaves and so on (UN-Habitat, 2011; WHO and UN-Habitat, 2016; WMO, 2003, 2016). In addition, the character of urban development (e.g. extensive paving) and the attendant emissions of pollutants have profoundly altered the natural setting and exacerbated natural hazards. These changes have led to a series of urban environmental problems, such as the urban heat island effect, the pollution of rivers and seas, air pollutant effects and air quality degradation that directly impact urban citizens (Mills, 2007; UN-Habitat, 2016a).

Moreover, the emissions of greenhouse gases (GHG) globally are concentrated in cities; although less than 3% of global land-cover can be described as urban, over 75% of human-sourced CO_2 arises from these places. Thus, there is an increasing international focus on cities as a nexus for many global problems, including urban environmental conditions, living quality and human well-being, the challenges and risks for human population health in cities, and global climate change (Cleugh et al., 2009; Corburn, 2015; UN-Habitat, 2016b; WHO, 2010; WHO and UN-Habitat, 2016). The 2016 United Nation's HABITAT-III conference adopted the *New Urban Agenda* to focus on sustainable urban development. According to the Quito Declaration, the agenda focuses on 'Sustainable Cities and Human Settlements for All', and one of the main objectives is to create urban resilience (UN-Habitat, 2016b) to mitigate extreme weather events. Various models and simulations can be applied to understand the future urban environment and to review the effects of past urbanisation. The UN also calls for using data science and analytics to achieve 17 Sustainable Development Goals (SDGs) – especially for SDG 11 sustainable cities and communities and SDG 13 climate action – by using satellite imagery and remote sensing technology.

Most cities operate like management systems that respond by mitigating the actions that cause undesirable changes and then adapting the system to cope with environmental hazards. Since cities have different capacities to respond based on their political cultures, economic base, socio-cultural make-up and so on, there needs to be a common language to describe urban landscapes globally. Members of the scientific community (e.g. climatologists, geographers, ecologists, environmental engineers and technologists) are looking for more standardised information on the urban form and function of cities so that they can devise solutions to aid architects, designers, and municipal governments (i.e. City Planning Departments, Public Works Departments, Zoning Boards, Transportation Offices) in developing evidence-based design guidelines for best practices of mitigation and adaptation to climate change (Cleugh et al., 2009). However, a consistent database on global cites at scales suitable for scientific inquiry and policy formulation does not exist (see Table 7.1 and Figure 7.1). It is quite often left to researchers and the general public to find the issues of data availability, accessibility and standardisation. Whilst there are several well-established worldwide land use land cover databases such as AVHRR, MODIS, and ESA Globalcover, almost all of them only include one built-up or urban category in their classification, which cannot represent the intra-urban morphological variation and differences.

The World Urban Database and Access Portal Tools (WUDAPT, www.wudapt.org) project has grown out of the need for better information on the urban form and function of cities globally (Bechtel et al., 2019; Ching et al., 2018). Figure 7.1 shows the comparison among well-established land use land cover databases and the newly developed WUDAPT data, which can tell more details of urban morphological characteristics, especially the built environment at the city scale. The analysis of such data is critical as it represents the only way to solve cause-and-effect urban problems before they become crises (Mills, 2005). Even the most recent report from the Intergovernmental Panel on Climate Change (IPCC) notes the dearth of such information on urban areas (IPCC, 2014).

Table 7.1 Problems and needs of urban data

Problems & needs	Facts
Urban data availability & accessibility	Few or no data sets available globally on cities, especially for rapidly growing places in economically developing countries and regions
Climate relevant urban data	No consistency in existing urban databases on spatial and temporal resolution and the variables to characterise the urban landscape
Data standardisation & harmonization	No standard classification to represent land use and land cover of cities and surroundings landscapes (e.g. densities, heights, functions and natural coverage for distinction)
Modelling needs	Need a global database that offers multiple properties on urban morphologies and landscapes (e.g. morphologies, geometrics, thermal / physical information, surface cover...)
Application needs	Should be applicable globally and transferable to each city (e.g. transdisciplinarity for scientific research, urban planning, disaster & risk management, health impact analysis and response, etc.)

Source: Oke (2006a, 2006b); Stewart and Oke (2009)

The WUDAPT project started in 2011 and is still under development. The current stage focuses on urban data extraction and development. Some studies have explored the potential possibility of different applications of the WUDAPT data. Cities are described using Local Climate Zones (LCZ) (Stewart and Oke, 2009), which are associated with a range of key urban climate model parameters and thus can serve as inputs to high resolution urban climate models (Bechtel et al., 2019). The LCZ scheme has 17 standard types, including two subsets: 10 built types and 7 land cover types. The WUDAPT is designed to be universal, simple and objective, with the aim to be part of a global protocol to derive information about form and function of cities. It uses free satellite images and the free software of Google Earth (GE) and SAGA-GIS and applies globally available Landsat satellite images as input data in urban climate and climate change modelling (Cai et al., 2016).

The urban data gap

From a climate perspective, we need to capture urban information on the spatial character of aspects of urban form and function (Batty, 2009; Raven et al., 2017). The *form* of a city can be described in terms of:

1. **Land surface cover,** including vegetation, soil, water and impervious/paved surface cover. Surface cover is related to, e.g., the ability of the landscape to intercept, store and dispose of water.
2. **Construction materials,** including concrete, steel, brick, glass, asphalt, etc. used to build the city. These fabrics have distinctive properties related to their ability to absorb, store and release heat.
3. **Urban morphology,** including the dimensions of individual buildings and their position in relation to each other. Morphology creates a complex three-dimensional geometry that controls access to daylight and sunshine, modifies the flow of air and traps pollutants at street level.

The *function* of a city describes the residential, commercial, industrial activities and all that these entail (i.e. water, materials, food and energy requirements). Together these capture the metabolism of a city, that is, the throughput of resources and wastes needed to sustain its operations and maintain and extend its buildings and infrastructure (Batty, 2009; Mills, 2005). The form and functions of a city are strongly correlated, for example, international evidence indicates that cities that are more densely built and occupied have lower transportation costs (Betanzo, 2007; Jaillon and Poon, 2008; Ng, 2012; Smith, 1984). High-density cities such as Singapore and Hong Kong, through compact and mixed-land use design conserve valuable land resources, reduce people's daily transport distance, and thus, the energy needed (Ng, 2009). Ideally, information on the form and functions of cities should be collected and analysed at *multiple temporal and spatial scales* to meet different needs and to tackle cause-and-effect problems (Raven et al., 2017).

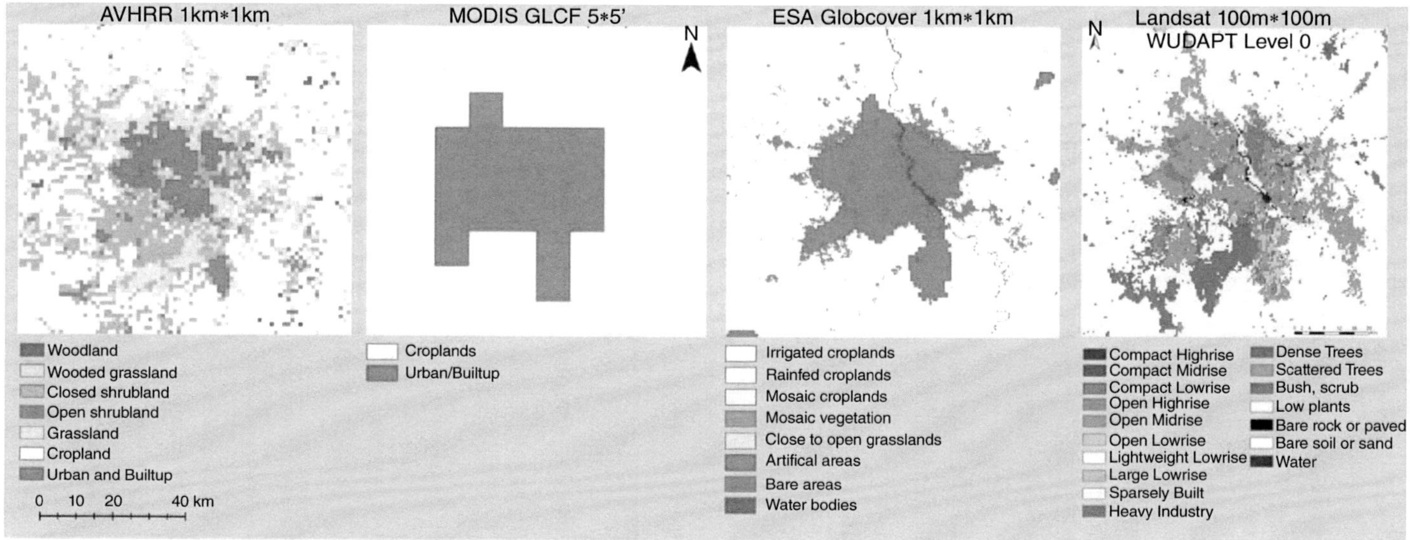

Figure 7.1 Land use challenges at city level. Currently available global scale data: Delhi, India

Source: ESA GlobCover 2009 project; Jackson et al. (2010)

World Urban Database and Access Portal Tools (WUDAPT)

The WUDAPT project seeks to acquire and store urban data using a common framework and to link these data to available methods for climate analysis and for current and what-if scenario development. Present global-scale urban data and land use at the city level is derived from global land cover data sets, e.g. AVHRR, MODIS GLCF, ESA GlobCover. Their resolutions mainly range from 1 km to 10 km and only have one land use category for urban/built-up areas, which may overestimate or underestimate urban forms and functions as well as not representing the actual situation of cities. Figure 1 shows the case of Delhi in India by using different existing land cover information and the final map in this sequence shows how the WUDAPT product improves spatial detail. There are other newly available global urban layers available from the German Aerospace Agency (DLR) (Global Urban Footprint) and the Joint Research Centre of the European Commission (Global Human Settlement Layer), but similarly, they show only one urban category of built-up areas.

The WUDAPT initiative is a global, bottom-up, self-organised effort that attempts to fill the data gaps to address the global challenges of sustainable cities and communities and to act as a guide to facilitate climate-based actions (Mills et al., 2015; See et al., 2015). It builds upon innovative approaches, employs community-based collaborations, and utilises existing and publicly available data where it exists (Ching et al., 2016). Given the urgent need for urban data to support climate research, WUDAPT has adopted a pragmatic approach to structuring these data according to the level of detail (see Figure 7.2). Table 7.2 shows the detail of the WUDAPT products, such as their coverage, data sources, resolution, applications and potential users.

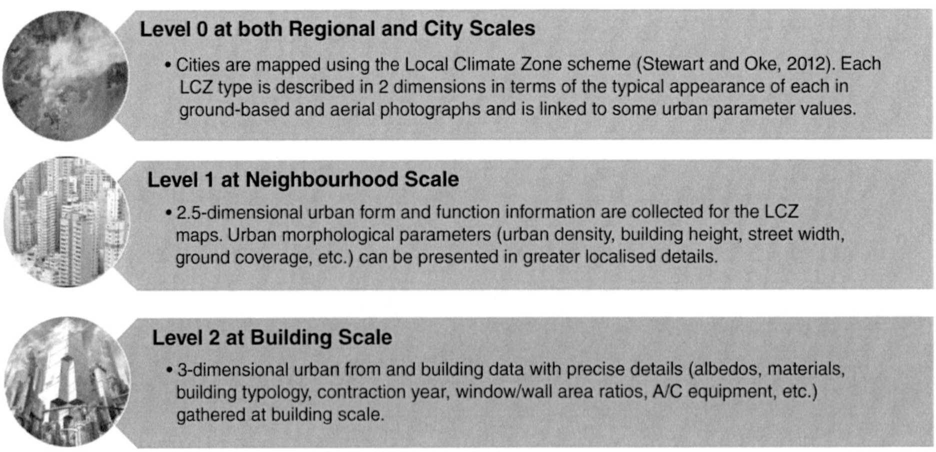

Figure 7.2 WUDAPT level 0, 1 and 2 products

Table 7.2 WUDAPT Levels 0, 1 and 2 data and their potential applications

WUDAPT data			
Product	**Level 0**	**Level 1**	**Level 2**
Coverage	Over 120 cities and regions	Data gathering methods testing based on building information	Any city by using our new 3-D mapping technology
Data source	Landsat + Google Earth	Landsat + Google Earth + local data & expert evaluation	World-view Stereo Data + Terra-SAR data
Resolution	100–500 m	100–500 m	2 m
Format	kml, tiff	GIS shapefiles	GIS shapefiles
Applications	Environment and Energy (Weather Research and Forecasting (WRF) modelling, urban heat island) Urban and Regional planning (population density)	• Environment and Energy (weather and climate, urban air flows, urban radiation, mean radiant temperature, urban energy consumption, air pollution, CO_2 and GHG emission) • Ecology (biodiversity) • Urban and Regional planning (master plan, land use plan, green master plan, new town plan)	• Environment and Energy (building energy cost) • Building and community design (visibility analysis, building development) • Disaster and risk management (flooding, heatwave) • Pedestrian and citizens' mobility (walkability, thermal comfort) • Public health (polluted areas)
Potential Users	Anyone Education institutions Researchers & Scientists	• Researchers & Scientists (Environment scientists, Climatologists, Meteorologists) • Engineering consultant companies and design firms • Education institutions • NGOs • Urban designers, town planners • City government (planning departments, transportation office, public works departments, zoning boards, etc.)	• Researchers & Scientists • Engineering consultant companies and design firms • NGOs • Education institutions • Architects, urban designers, town planners • City government (planning departments, transportation office, public works departments, zoning boards, infrastructure facilities, etc.)

An important component of WUDAPT is the development of portal tools which allow data to be extracted in formats suitable for climatic analysis. This includes sophisticated models such as the Weather Research and Forecasting (WRF) model (Skamarock et al., 2008), a widely employed community-based model that is capable of simulating all the parameters of climate and, depending on the setup, air quality and climate change scenarios (Ching et al., 2018). In fact, the major impediment to applying these models – which represent decades of research knowledge – is the lack of data.

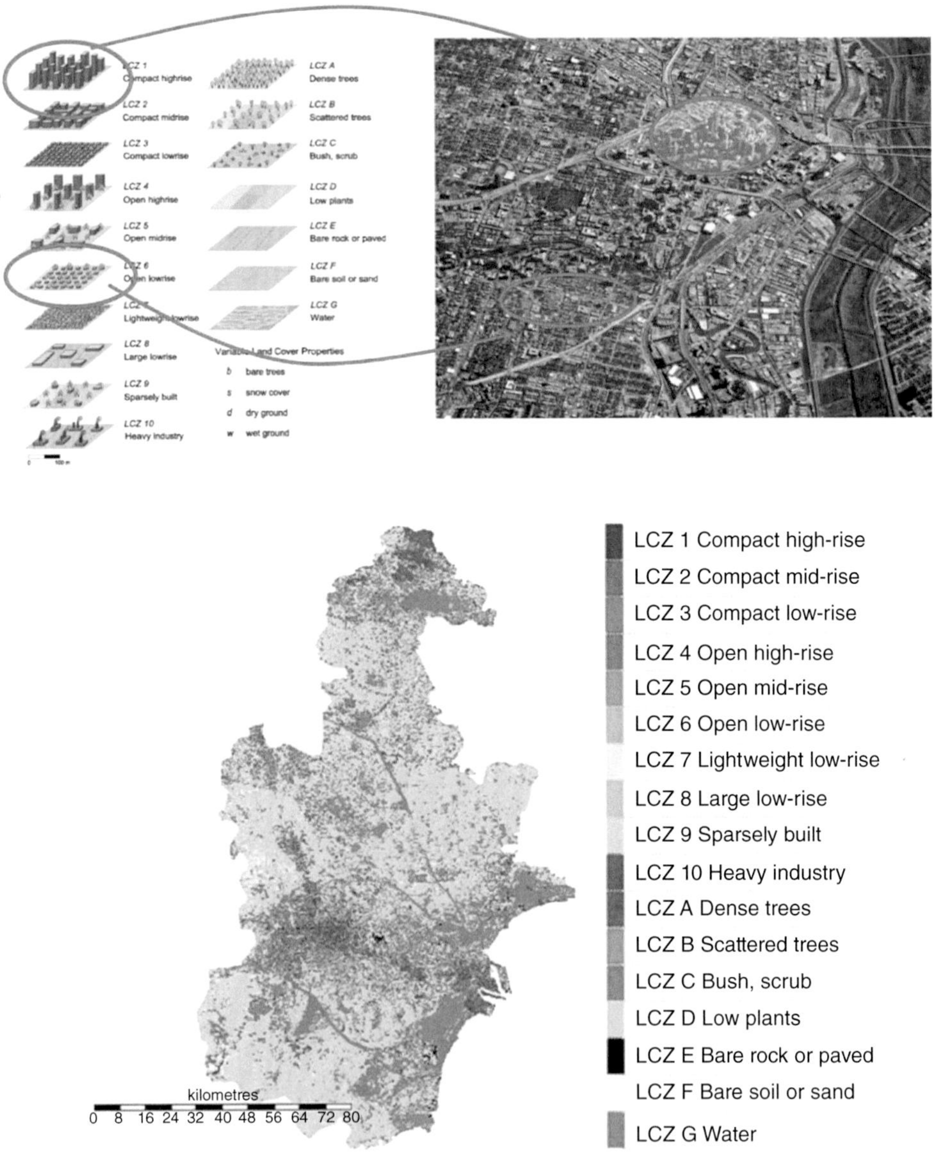

Figure 7.3 Generating L0 data. The top image shows how parts of the urban landscape are identified according to LCZ type. These data are used to automatically classify an entire urban region into common LCZ types (lower map)

The WUDAPT approach is underpinned by a community-based approach and relies on freely available data and tools. The lowest level of detail (L0) consists of decomposing urban (and surrounding landscapes) into common Local Climate Zone (LCZ) types using local expertise, Landsat remote sensing data and software tools (Figure 7.3) (Ching et al., 2018). The net result is an LCZ map of an urban region where each LCZ type is associated with a universal range of values that describe aspects of urban form and function. Currently, L0 data are available for more than 120 city regions across all continents. L0 data can be used as a sampling framework to gather more detailed information on cities (L1 data).

Level 2 (L2) data provide complete coverage of the urban landscape and include information on individual urban elements (e.g. buildings, trees), and Figure 7.4 provides an example of 3-dimentional urban morphology (Xu et al., 2017). Ideally, this information would be supplemented by information on the occupation patterns of buildings (commercial, residential, mixed use) and on their material composition. While the morphology can be acquired through remote sensing, other properties require a more holistic approach that may need traditional census data and crowd-sourced information.

Figure 7.4 Level 2 product: 3D urban morphological data of Kowloon Peninsula, Hong Kong, based on innovative technology (under patent) which can extract and detect 3-dimensional urban morphological data by adopting multi-source satellite images. The overall accuracy of developed level 1 data can reach to 70–90%. This new technology has a low cost and a large spatial coverage, which can be updated efficiently

Collaboration and application: the Pearl River Delta (PRD) megalopolis project

Since its inception, the WUDAPT project has established working relationships with a number of organisations. We have a close association with the International Association for Urban Climate (IAUC), which has a diverse (e.g. architects, designers and planners, climatologists and meteorologists) and globally distributed membership with interests in urban environmental issues, weather/climate prediction, air pollution and risk management. Many of the local experts that we draw upon to provide city-specific information are IAUC members. In addition, WUDAPT has established strong collaborations with the Human Planet Initiative, the Global Carbon Project (GCP) and the Digital Belt and Road Initiative (DBAR). With funding from the Research Grant Council of Hong Kong, an extensive study of the Pearl River Delta has been undertaken using the WUDAPT methodology. Also, an informal collaboration with the World Meteorological Organization (WMO) is underway to provide supporting infrastructure to support their recent Urban Mandate for providing urban services.

The Greater Pearl River Delta (GPRD) includes the Hong Kong Special Administrative Region (HKSAR), the Macao Special Administrative Region and the Pearl River Delta (PRD) Economic Zone in Guangdong Province. The GPRD region is located at the mouth of the Pearl River estuary, where the river enters the South China Sea. The PRD Economic Zone, as specified by Guangdong Province, includes nine municipalities, namely Guangzhou, Shenzhen, Dongguan, Foshan, Jiangmen, Zhongshan, Zhuhai, and the urban areas of Huizhou and Zhaoqing (HK Government, 2015) (Figure 7.5). It is one of the most densely urbanised regions in the world. With China's remarkable economic growth since the beginning of the country's reform period in 1978, urbanised areas have increased by 8.47% annually, rising from 1,068.7 km² in 1979 to 4,617.16 km² in 2008. The rapidly

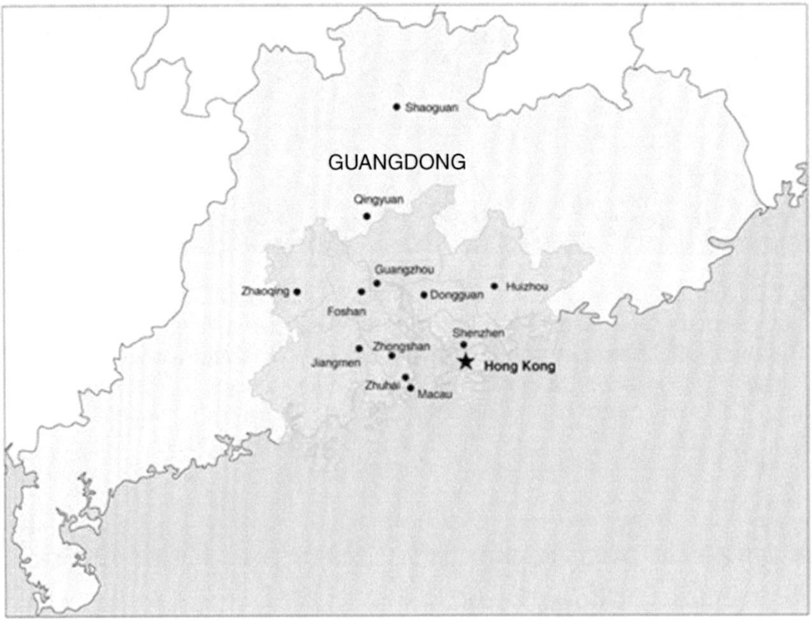

Figure 7.5 Map of the PRD region

expanding urban areas have emerged into one large region and formed a megaregion and urban agglomeration. Given these dramatic land use changes, it is not surprising that the GPRD has experienced noticeable regional climate changes. The Guangdong Meteorological Administration reported that the GPRD witnessed great temperature change of averaging +0.3 °C every 10 years (Guangdong Government, 2007). Guangzhou is located in the south of China with a total area of 7,434 km^2 (He et al., 2010). It is one of the most urbanised cities in GPRD and megacities not only in China, but also on earth. During its rapid urban expansion, the previously dispersed built-up areas have been integrated to a large extent and spread into the adjacent cities like Foshan and Dongguan (Mu et al., 2007). In addition, Guangzhou has a high people, of about 8.5 million people, greatly increasing the anthropogenic heat emissions (Xu, 2012). Therefore, the urban heat island (UHI) phenomenon is a significant, moreover immediate, problem of Guangzhou and the GPRD that requires research investigation.

Despite being the most densely populated and urbanised region in the world, urban expansion and intensive development in the PRD will continue according to *The PRD Region Reform and Development Plan (2008–2020)*. In the past urbanisation process, the whole region has created unique compact and dense urban morphological forms. Its local urban climatic conditions have also been inevitably and gradually changed. Thus, understanding the urban development impact on local urban climatic changes from the 1980s to the 2010s will allow planners to gain insights into the future.

Data

A proper land-surface model is particularly important in the simulation, especially because the objective of this study is to analyse meteorological changes due to rapid urbanisation in the PRD region. Landsat 8 Level 1 images of the PRD with the resolution of 30m were downloaded from the USGS (U.S. Geological Survey). Urban morphology information, including the building and land cover was derived from Google Earth (GE). Figure 7.6 shows the training samples in the PRD region. They are evenly distributed to get a better classification result. There are in total 17 categories defined for the LCZ classification and 10 represent urban categories; but the simulation model used in this study does not currently support the LCZ classification as input data, so it is necessary to remap these data in the USGS land-cover classification. However, the USGS land-cover classification has only one land use code to represent urban area, which is not representative in this study due to the heterogeneous nature of urban areas in the PRD region. As such, two more urban categories are defined to further differentiate urban areas into areas with low-rise, mid-rise, and high-rise buildings. After modifying the USGS land-cover classification, the categories in the LCZ classification are then remapped to follow the classification of USGS land-cover. The remapping results are shown in Figure 7.7 (only urban categories are shown). Moreover, a portion of the land use data was corrected manually via subjective analysis, especially when urban areas appeared on the sea. Such false classifications may occur due to poor data quality (Wang et al., 2018).

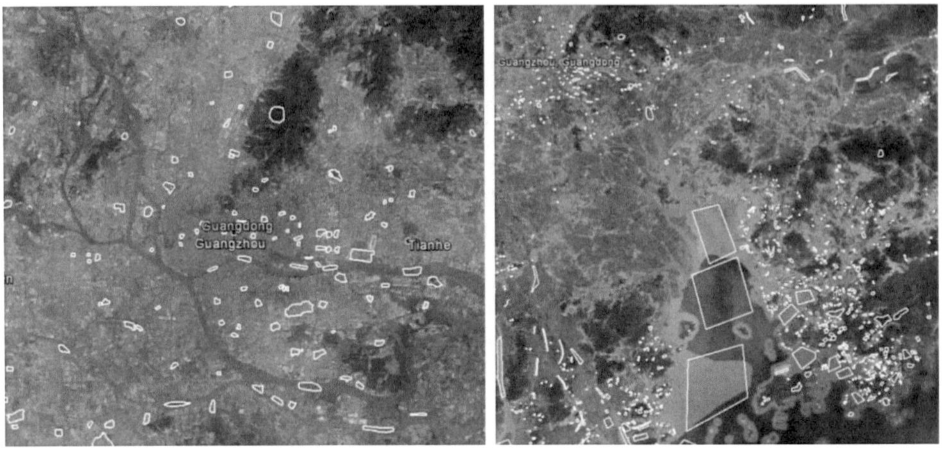

Figure 7.6 (a) Screenshots of training samples of Guangzhou from GE; (b) Screenshots of training samples of GPRD from GE

Source: Google Earth, Data collected on 16 August 2016

Main research steps

This study firstly collects historical Landsat images and detects land use land cover and urban morphological data representing the 1990s, 2000s and 2010s. This classification standard follows the guidelines of land use classification from WUDAPT (Bechtel et al., 2019). Secondly, referring to the USGS land use category, the WUDAPT level 0 product could be re-classified. For the urban category, three sub-categories of high-roughness, mid-roughness, and low-roughness are developed and their corresponding settings in the WRF simulation model are updated and redefined. Thirdly, the WRF simulation with the default setup for the year 2010 is run for the summer period and its results serve as the benchmark. Two new WRF simulations, using the WUDAPT products as input data, are conducted and their results are then compared with the first benchmark to establish the pure urbanisation (1980–2010) impact on local climate conditions in the summertime.

In this study, four simulation domains are configured (Figures 7.8a and 7.8b). The horizontal grid resolution for domains 1, 2, 3 and 4 are 27km, 9km, 3km and 1km, respectively. The fourth domain is the area of interest for this study, where there are numerous mountains with height ranging between 150m and 750m located at Jiangmen, Hong Kong, Shenzhen, and Huizhou respectively. The estuary of the Pearl River is a low lying plain. The asymmetric convective model (Pleim, 2007), which is a non-local planetary boundary layer scheme, was chosen as the physics scheme for the WRF model.

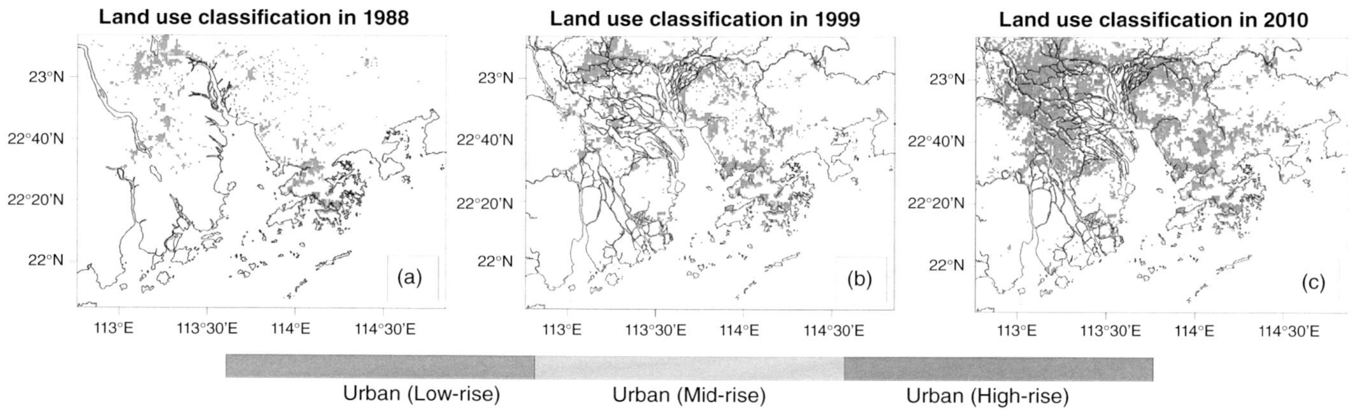

Figure 7.7 Classification of land use following USGS landcover classification in the PRD region (after manual correction; only urban categories are shown) in (a) 1988, (b) 1999, and (c) 2010

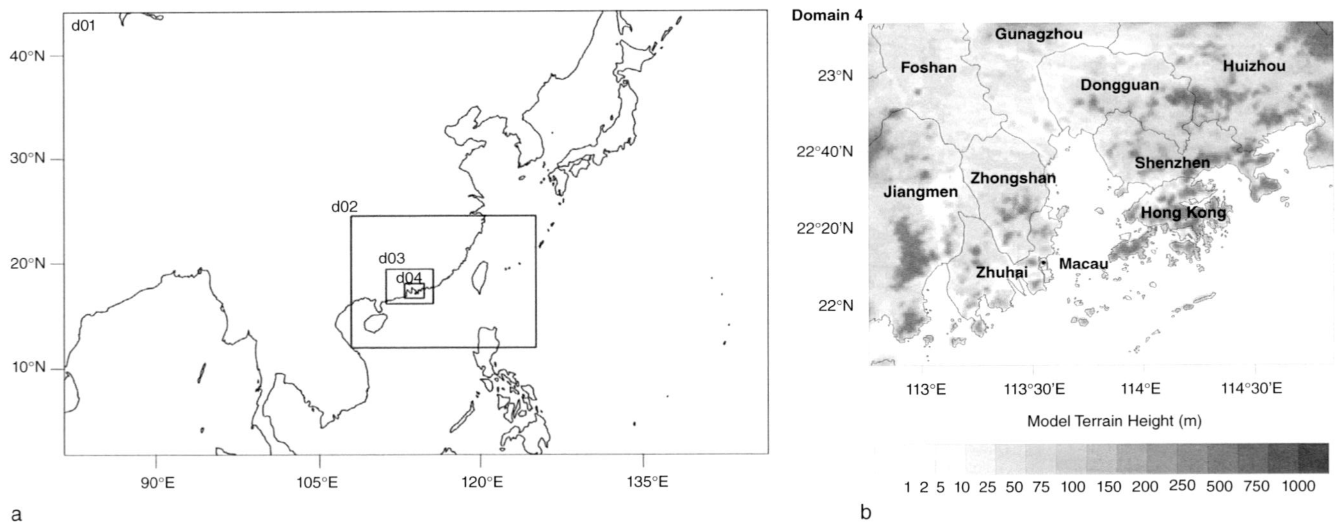

Figure 7.8 a) Four simulation domains set up in the WRF model; b) Fourth domain (PRD region) with major cities marked and contour plot of terrain height

In this study, the volumetric heat capacities and roughness lengths of the urban land uses are updated accordingly after changing the land-cover classification. The volumetric heat capacity of low-rise, mid-rise, and high-rise urban areas are consistent with the concrete materials used in buildings. Another changed parameter is the roughness length. The roughness lengths of the low-rise, mid-rise, and high-rise urban land use are set at 0.5m, 1.0m and 1.8m, respectively. These values refer to the study by Grimmond and Oke (1999) on the aerodynamic properties of urban areas. The techniques of changing urban parameters under the Noah-LSM have also been used to capture the urban effect (Liu et al., 2006).

Since the study period includes the summer season (June to August), to avoid deviation from the observation data due to the long simulation time, the simulation period is divided into four-day WRF simulations and the configuration from the first day of the simulation is used as the model spin-up time for initialisation. In the simulation, the initial and boundary conditions are provided for the WRF model from the National Centers for Environmental Prediction Final Operation Global Analysis data by the National Center for Atmospheric Research. Because the initial and boundary conditions for the different simulations with 1988, 1999 and 2010 land use data are all the same, the changes could be interpreted as purely the effects of urbanisation and its associated interaction with other physical parameters, such as the southern warm and moist oceanic air mass and solar irradiance, on the local climate.

Results and analysis

This section focuses on seasonal diurnal variation. For ease of analysis, 'day' and 'night' are defined as 1400 to 1600 local standard time (LST) (UTC + 8) and 0200 to 0400 LST, respectively, because urban effects such as the UHI effect are common at these times. The data presented in this section are averaged throughout the aforementioned time periods corresponding to the simulation years (1988, 1999 and 2010), with 1988 chosen as the reference year to evaluate changes since the beginning stages of urbanisation in the PRD region. Moreover, using the averaged data, three fundamental climatic variables were investigated: 2-m air temperature, 10-m wind speed, and the heat index developed by the U.S. National Weather Service.

Figures 7.9a and 7.10a show that parts of Guangzhou and Foshan were already the hottest areas in 1988 and became the warm centre because of pre-existing urban areas. In 1999 and 2010, more areas saw temperature increases from urbanisation, and some areas saw a 2 °C increase from 1988 (Figure 7.9e). The warm centre not only intensified, but also expanded and covered considerable areas in the western part of the PRD. Similar results were found for nighttime temperatures. Figures 7.10d and 7.10e also show that the urbanised areas experienced the greatest increase in temperature at night. Apart from the urbanised areas, their surrounding rural areas also experienced a slight temperature rise, which can be attributed to the heat diffused from urban areas.

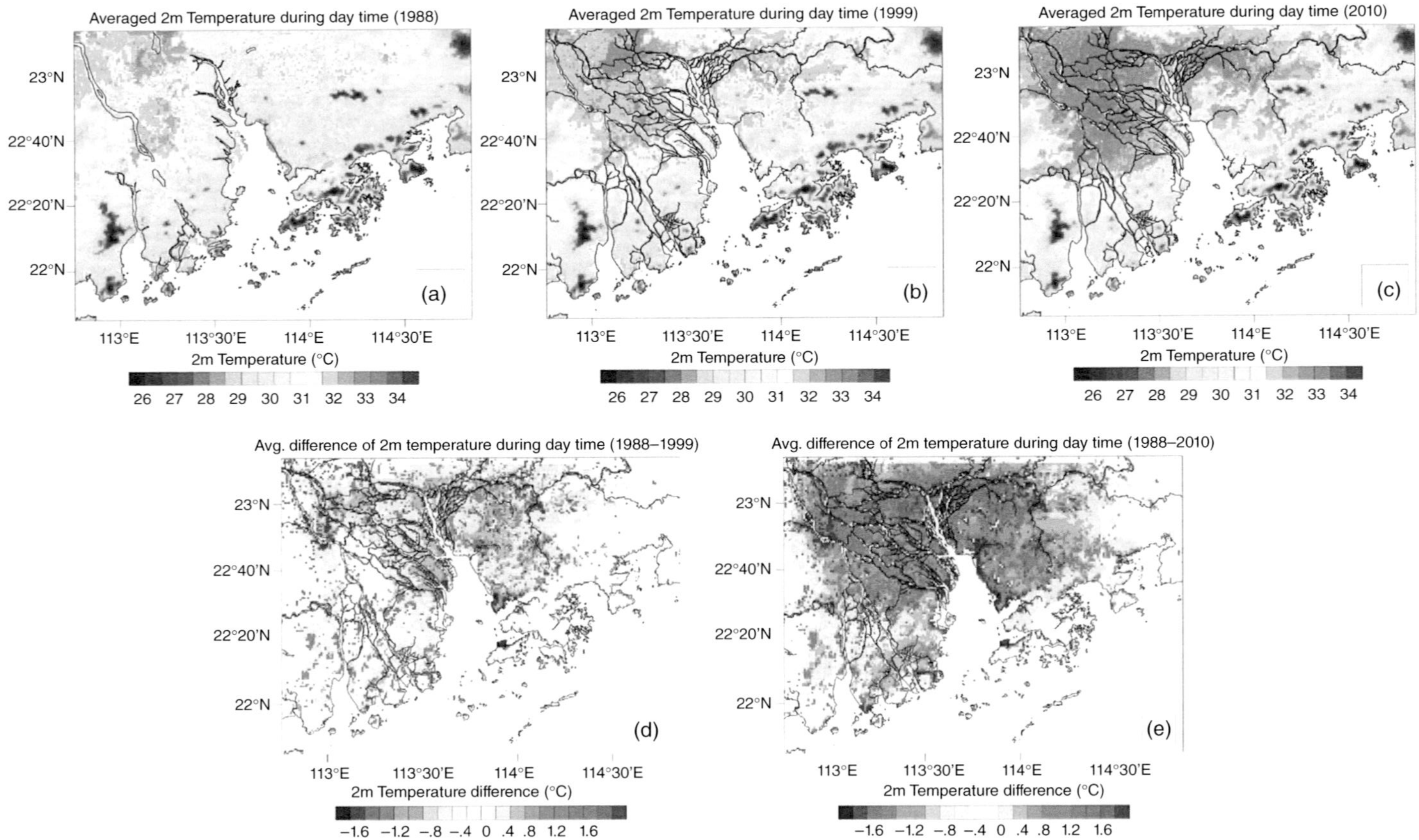

Figure 7.9 Average daytime 2m temperature over land in (a) 1988, (b) 1999, (c) and 2010 and (d–e) their difference

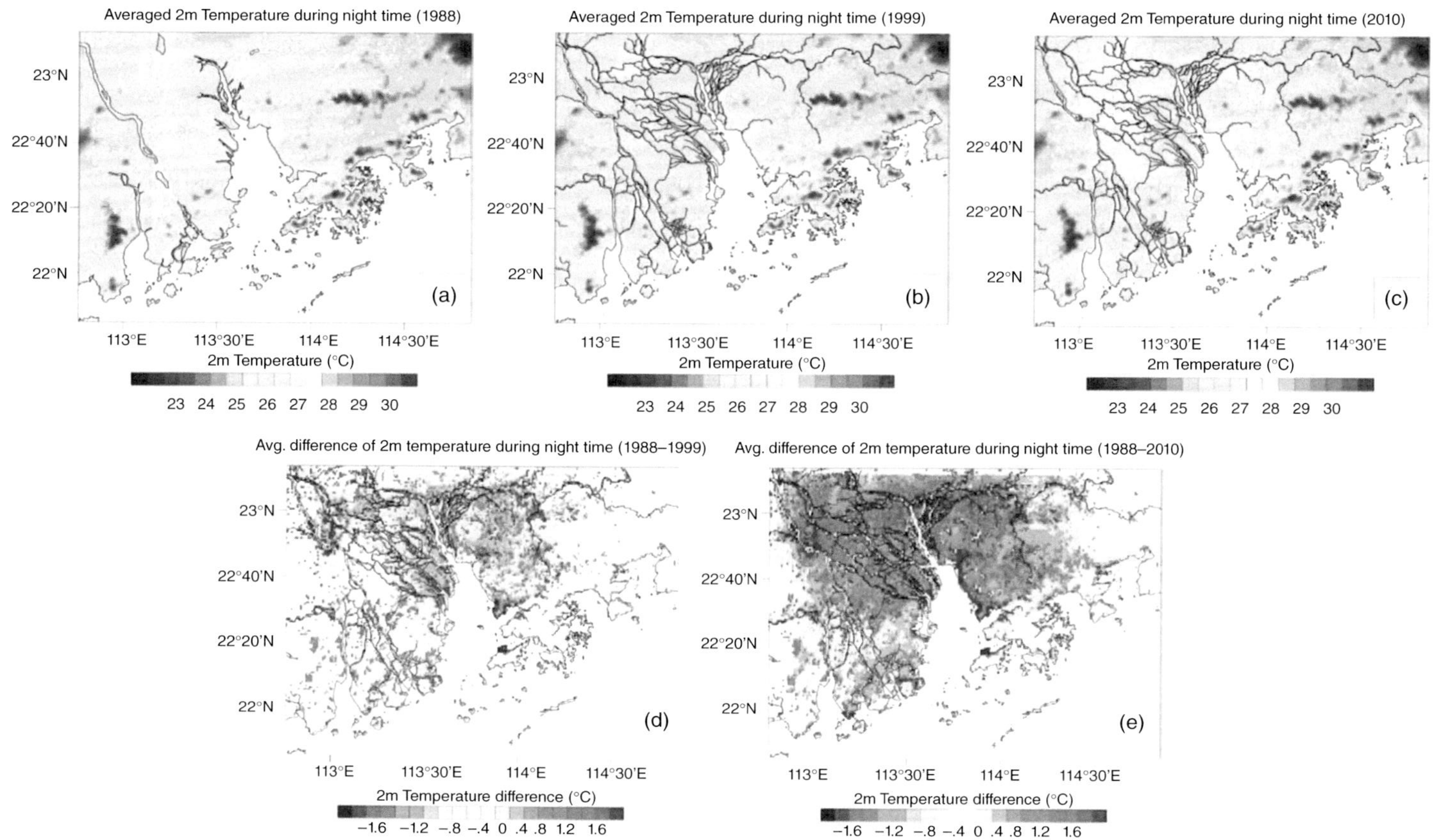

Figure 7.10 Average nighttime 2m temperature over land in (a) 1988, (b) 1999, (c) and 2010 and (d–e) their difference

The higher temperature in urban areas can be explained by the change of physical characteristics due to urbanisation. First, together with the reduced surface albedo due to urbanisation, buildings in urban areas allow a larger amount of energy to be absorbed during the day than those in rural areas. The absorbed energy is released at night, so higher temperatures are maintained throughout the night (Taha, 1997). Second, due to the reduction of soil moisture and vegetation in urban areas, there is a general increase in sensible heat flux and decrease in latent heat flux during the day to maintain the surface energy balance (Rizwan et al., 2008; Taha, 1997). When compared with the year 1988, most of the urbanised areas in 2010 have an increase of around $300Wm^{-2}$ in sensible heat flux and the same amount of decrease in latent heat flux; while there are no significant changes for most of the rural areas as they remain undeveloped. Overall, as the sensible heat flux in urban areas is strengthened, the temperature is also increased correspondingly.

The differences in the seasonal average of wind speed during the day and at night are generally weakened when compared to 1988, as seen in Figure 7.11. The result is consistent with the observation data from Hong Kong, which suggests that the wind speed in urban areas is declining due to urbanisation (HKO, 2016). As the urbanisation process will result in more buildings in urban areas, the corresponding surface roughness is increased. Consequently, the magnitude of wind speed is reduced.

The heat index aims to model the human-perceived equivalent temperature by accounting for various meteorological variables to evaluate the associated risks. The heat index model developed by the U.S. National Weather Service (Rothfusz, 1990) was chosen for this study. It uses an empirical formula associated with 2m relative humidity (R) and 2m air temperature (T) to determine the Human-perceived equivalent temperature in shaded areas and light wind conditions, which is shown below.

$$\text{Heat Index} = c_1 + c_2 T + c_3 R + c_4 TR + c_5 T^2 + c_6 R^2 + c_7 T^2R + c_8 TR^2 + c_9 T^2R^2 \qquad (7.1)$$

We adopted the above formula to estimate the potential heat stress on the human body which could be caused by the changing climate. In the formula, c is a constant for $1 \leq i \leq 9$. This formula is obtained by a multivariate fit to a model of the human body. The value calculated in the formula can be placed into four categories, as summarised in Table 7.3.

Because most areas in the PRD fall into the 'Extreme Caution' category (> 32 °C), the focus is on the comparison of the heat index. Figures 7.12a and 7.12b show that the increase in heat index is widespread during the day in both urban and rural areas

Table 7.3 Classification of heat index value

Classification	Range (°C)	Effect on the body
Caution	27 to 32	Fatigue possible with prolonged exposure and/or physical activity
Extreme caution	32 to 41	Heat stroke, heat cramps, or heat exhaustion possible with prolonged exposure and/or physical activity
Danger	41 to 54	Heat cramps or heat exhaustion likely, and heat stroke possible with prolonged exposure and/or physical activity
Extreme danger	Over 54	Heat stroke highly likely

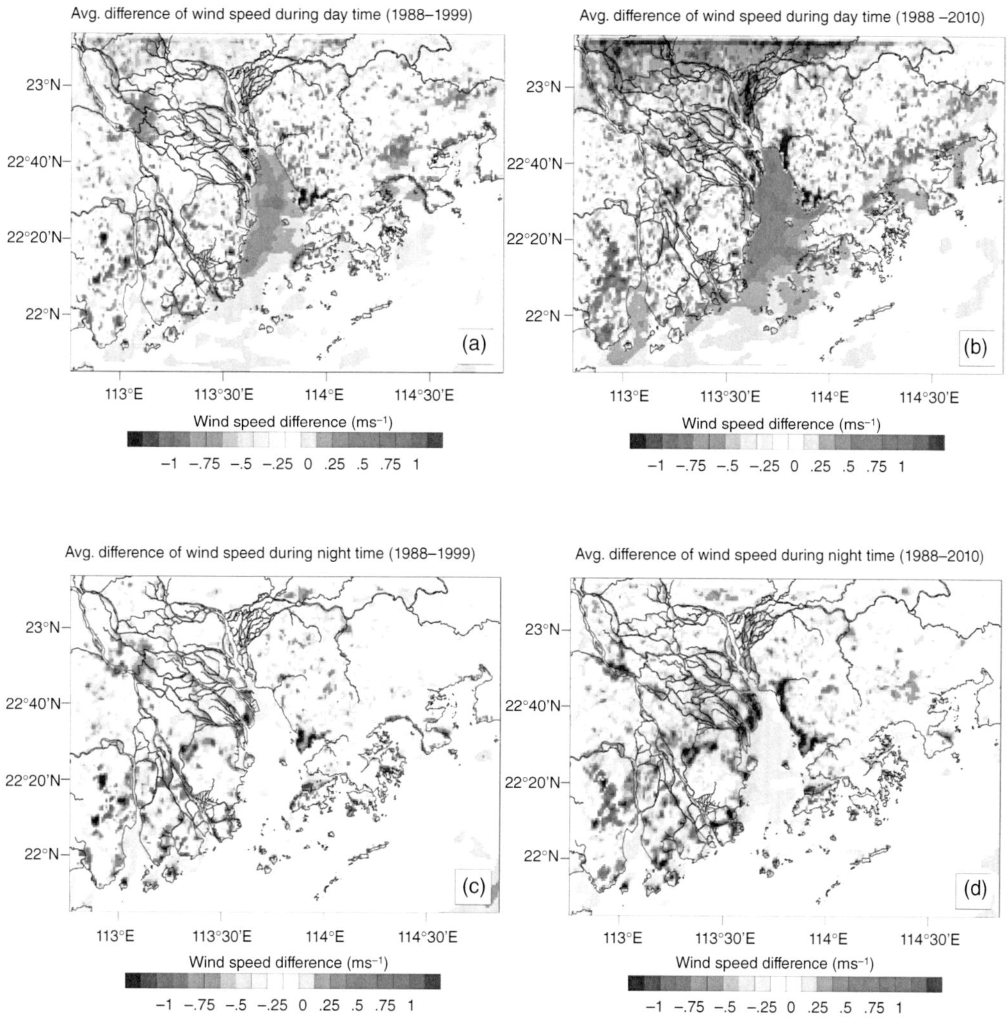

Figure 7.11 Average difference of 10m wind speed during day (a–b) and at night (c–d), respectively, in different years

near the coast, possibly because of intensified sea breezes and the widely urbanised area. The former facilitates the transportation of moisture from the sea to the coastal area and increases the relative humidity, whereas the latter increases the temperature. Both effects cause a significant increase in the heat index of coastal areas because the heat index depends on relative humidity and temperature. At night, however (Figures 7.12c and 7.12d), the difference in the heat index of coastal areas is insignificant because the sea breeze changes less near the coast (Figures 7.11c and 7.11d). In contrast to coastal areas, inland areas generally have a considerable increase in the heat index, which is enhanced at night. Some areas can even reach a 2.5 °C increase in the heat index, particularly the eastern part of Shenzhen and the northwestern part of the PRD (as shown in Figure 7.12d).

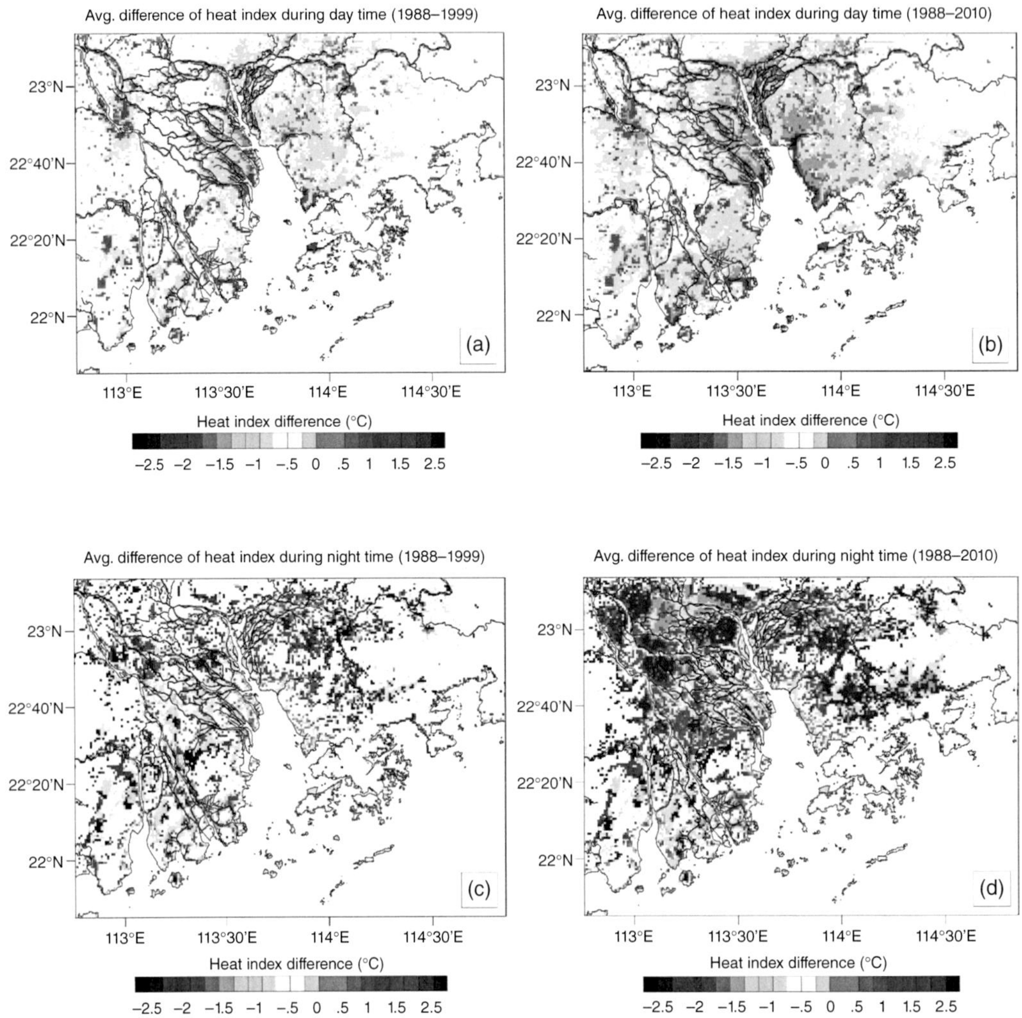

Figure 7.12 Average difference of heat index over land during day (a–b) and at night (c–d) in different years

Conclusion

The PRD region is the youngest and biggest mega urban region in the world. This study demonstrates the climatological changes caused by rapid urbanisation in the PRD region between 1988 and 2010. Land use data for different decades were created by applying machine learning techniques to classify the land use based on satellite images. The simulations were then performed with the WRF model with the corresponding land use data, while maintaining the same initial and boundary conditions to allow the changes to be investigated. After completion of the simulation, climatic variables like the 2m air temperature, 10m wind speed, and heat index were investigated through spatial cross-comparison and statistical analysis. It is expected that the results of this study could serve as a reference for the local planners, especially with the future development of the Guangzhou–Hong Kong–Macao bay area by the Chinese government. Through understanding of the impact of urbanisation, governors can plan correspondingly to alleviate its negative impact, for example, the UHI effect.

The lessons learned from this study can provide local policymakers and planners with a quantitative understanding of the urban development of the PRD region. It also can serve as a useful reference for developing countries in sub-tropical climate regions which are also facing fast urbanisation. By capturing detailed urban features and their impact with the help of WUDAPT data, the findings can inform documents such as *The environmental performance assessment guide of urban ecological development of China (Trial Version)* and *The PRD Region Reform and Development Plan (2008–2020)* (Guangdong Government, 2008).

The WUDAPT initiative has attracted a fast growing community to work together on urban data development and its application into urban climate and urban meteorology related studies. Now more than 150 cities' data are available online. This global initiative would not only develop a new set of standardised urban data, but also encourage cross-disciplinary collaborations on urban morphology, urban climatology and public health impact.

Acknowledgements

This work was supported by NSFC/RGC Grant N_HKUST631/05, NSFC-FD Grant U1033001, RGC Grant 16303416, GRF Grant 14611015 and 14643816. This study is also supported by the Vice-Chancellor's One-off Discretionary Fund of The Chinese University of Hong Kong. This chapter is further developed based on Tse et al. (2018).

References

Batty, M. (2009) 'Cities as complex systems: Scaling, interaction, networks, dynamics and urban morphologies', in R.A. Meyers (ed.), *Encyclopedia of Complexity and Systems Science*. Berlin: Springer. pp. 1041–71.

Bechtel, B., Alexander, P.J., Beck, C., Böhner, J., Brousse, O., Ching, J. [...] and Xu, Y. (2019) 'Generating WUDAPT Level 0 data – Current status of production and evaluation', *Urban Climate*, 27: 24–45. doi.org/10.1016/j.uclim.2018.10.001

Betanzo, M. (2007) 'Pros and cons of high density urban environment', *Build*, April/May, 39–40. https://www.buildmagazine.org.nz/assets/PDF/B99-39-UrbanEnviro.pdf (accessed January 1, 2021).

Cai, M., Ren, C., Xu, Y., Dai, W. and Wang, X.M. (2016) 'Local climate zone study for sustainable megacities development by using improved WUDAPT methodology – A case study in Guangzhou', *Procedia Environmental Sciences*, 36: 82–9. doi.org/10.1016/j.proenv.2016.09.017

Ching, J., Mills, G., See, L., Benchtel, B., Feddema, J., Stewart, I., [...] and Masson, V. (2016) 'WUDAPT (World Urban Database and Access Portal Tools): An international collaborative project for climate relevant physical geography data for the world's cities',

The 96th American Meteorological Scoiety (AMS) Annual Meeting. New Orleans, LA: American Meteorological Society.

Ching, J., Mills, G., Bechtel, B., See, L., Feddema, J., Wang, X. [...] and Theeuwes, N. (2018) 'WUDAPT: An urban weather, climate and environmental modeling infrastructure for the Anthropocene', *Bulletin of the American Meteorological Society*, 99 (9): 1907–24. doi.org/10.1175/bams-d-16-0236.1

Cleugh, H., Emmanuel, R., Endlicher, W., Erell, E., Mcgranahan, G., Mills, G., Ng, E., Nickson, A., Rosenthal, J. and Steemer, K. (2009) 'Climate and sustainable cities: Climate information for improved planning and management of mega cities (needs and capabilities perspectives)', *World Climate Conference-3*. Geneva, Switzerland.

Corburn, J. (2015) 'Urban inequities, population health and spatial planning', in H. Barton, S. Burgess and M. Grant (eds), *The Routledge Handbook Of Planning For Health and Well-Being: Shaping a Sustainable and Healthy Future*. London: Routledge. pp. 37–47.

ESA (2010) GlobCover 2009 project. Release on 21 December 2010, available from http://due.esrin.esa.int/page_globcover.php

Grimmond, C.S.B. and Oke, T.R. (1999) 'Aerodynamic properties of urban areas derived from analysis of surface form', *Journal of Applies Meteorology,* 38: 1262–92. doi.org/10.1175/1520-0450(1999)038<1262:APOUAD>2.0.CO;2

Guangdong Government (2007) *Assessment Report on Climate Change in Guangdong 2007*. http://www.gdsafety.gov.cn/

Guangdong Government (2008) *The Plan for the Reform and Development of the PRD (2008–2020)*. http://www.gdsafety.gov.cn/

He, C., Lei, Y. and Li, G. (2010) *Urban Development and Climate Change in the Pearl River Delta*. Cambridge, MA: Lincoln Institute of Land Policy.

HK Government (2015) *An Overview of the Greater Pearl River Delta*. https://www.cmab.gov.hk/tc/images/issues/ar0405/F_GPRD_Overview.pdf (accessed January 1, 2021).

Hong Kong Observatory (HKO) (2016) 'Annual average of 12-hourly 10-minute mean wind speed of King's Park and Waglan Island (1968–2015)', *Hong Kong Observatory*. https://www.hko.gov.hk/en/climate_change/obs_hk_wind.htm (accessed January 1, 2021).

IPCC (2014) 'Summary for policymakers', in C.B. Field, V.R. Barros, D.J. Dokken, K.J. Mach, M.D. Mastrandrea, T.E. Bilir [...] and L.L. White (eds), *Climate Change 2014. Impacts, Adaptation and Vulnerability. Part A: Global and Sectoral Aspects. Contribution of Working Group II to the Fifth Assessment Report of the Intergovernmental Panel on Climate Change*. Cambridge: Cambridge University Press. pp. 1–32. Available from https://www.ipcc.ch/report/ar5/wg2/ (accessed January 1, 2021).

Jackson, T.L., Feddema, J.J., Oleson, K.W., Bonan, G.B. and Bauer, J.T. (2010) 'Parameterization of urban characteristics for global climate modeling', *Annals of the Association of American Geographers,* 100 (4): 848–65.

Jaillon, L. and Poon, C.S. (2008) 'Sustainable construction aspects of using prefabrication in dense urban environment: A Hong Kong case study', *Construction Management and Economics*, 26 (9): 953–66.

Liu, Y., Chen, F., Warner, T. and Basara, J. (2006) 'Verification of a mesoscale data-assimilation and forecasting system for the Oklahoma City area during the Joint Urban 2003 field project', *Journal of Applied Meteorology and Climatology* 45 (7): 912–29. https://doi.org/10.1175/JAM2383.1

Mills, G. (2005) 'Urban form, function and climate', *Heat Island Effect*. Webcast Presentations: U.S. Environmental Protection Agency. https://web.archive.org/web/20090927084456/www.epa.gov/hiri/resources/pdf/GMills4.pdf (accessed January 1, 2021).

Mills, G. (2007) 'Cities as agents of global change', *International Journal of Climatology*, 27: 1849–57.

Mills, G., Ching, J., See, L., Bechtel, B., Feddema, J., Masson, V. [...] and Steeneveld, G.-J. (2015) 'Introduction to the WUDAPT Project', The 9th International Conference on Urban Climate. Toulouse, France.

Mu, F., Zhang, Z., Tan, W. and Liu, B. (2007) 'Analysis on the spatial-temporal characteristics of Guangzhou City's spatial morphologic evolution', *Geo-Information Science*, 9 (5): 94–8.

Ng, E. (ed.) (2009) *Designing High-Density Cities for Social and Environmental Sustainability*. London: Earthscan.

Ng, E. (2012) 'Towards planning and practical understanding of the need for meteorological and climatic information in the design of high-density cities: A case-based study of Hong Kong', *International Journal of Climatology*, 32 (4), 582–98. doi.org/10.1002/joc.2292

Oke, T.R. (2006a) 'Initial guidance to obtain representative meteorological observations at urban sites', *Instruments and Observing Methods Report No. 81, WMO/TD-No. 1250*. Geneva: World Meteorological Organization.

Oke, T.R. (2006b) 'Towards better scientific communication in urban climate', *Theoretical and Applied Climatology*, 84: 179–90.

Pleim, J.E. (2007) 'A combined local and nonlocal closure model for the atmospheric boundary layer. Part I: Model description and testing', *Journal of Applied Meteorology and Climatology*, 46 (9): 1383–95. https://doi.org/10.1175/JAM2539.1

Raven, J., Brian, S.J., Gerald, M., Towers, J., Katzschner, L., Leone, M.F., Gaborit, P., Georgescu, M. and Hariri, M. (2017) 'Urban planning and urban design', in C. Rosenzweig, W. Solecki, P. Romero-Lankao, S. Dhakal and S. Ali-Ibrahim (eds), *Climate Change and Cities: Second Assessment Report of the Urban Climate Change Research Network*. Cambridge: Cambridge University Press. pp. 139–72.

Rizwan, A.M., Dennis, L.Y.C. and Liu, C. (2008) 'A review on the generation, determination and mitigation of Urban Heat Island', *Journal of Environmental Science*, 20 (1): 120–128. Doi: 10.1016/S1001-0742(08)60019-4

Rothfusz, L.P. (1990) 'The heat index "equation" (or, more than you ever wanted to know about heat index)', *National Weather Service Technical Attachment (SR 90–23)*. National Weather Service, Fort Worth. https://wonder.cdc.gov/wonder/help/Climate/ta_htindx.PDF (accessed January 1, 2021).

See, L., Mills, G. and Ching, J. (2015) 'Climate modelling: Community initiative tackles urban heat', *Nature*, 526 (7571): 43.

Skamarock, W.C., Klemp, J.B., Dudhia, J., Gill, D.O., Barker, D.M., Duda, M.G., Huang, X.-Y., Wang, W. and Powers, J.G. (2008) *A description of the advanced research WRF version 3* (NCAR Tech. Note NCAR/TN-475+STR). https://opensky.ucar.edu/islandora/object/technotes:500 (accessed January 1, 2021).

Smith, W.S. (1984) 'Mass transport for high-rise high-density living', *Journal of Transportation Engineering*, 110 (6): 521–35.

Stewart, I.D. and Oke, T. (2009) 'A new classification system for urban climate sites', *Bulletin of the American Meteorological Society*, 90: 922–3.

Stewart, I.D. and Oke, T.R. (2012) 'Local Climate Zones for Urban Temperature Studies', *Bulletin of the American Meteorological Society*, 93 (12): 1879–1900. Doi: 10.1175/bams-d-11-00019.1

Taha, H. (1997) 'Urban climates and heat islands: albedo, evapotranspiration, and anthropogenic heat', *Energy Build*, 25: 99–103. Doi: 10.1016/S0378-7788(96)00999-1

Tse, J.W.P., Yeung, P.S., Fung, J.C.-H., Ren, C., Wang, R., Wong, M.M.-F. and Meng, C.A.I. (2018) 'Investigation of the meteorological effects of urbanization in recent decades: A case study of major cities in Pearl River Delta', *Urban Climate*, 26: 174–87.doi.org/10.1016/j.uclim.2018.08.007

United Nations-Habitat (UN-Habitat) (2011) *Global Report on Human Settlements 2011: Cities and Climate Change*, London: Earthscan. Available from https://unhabitat.org/global-report-on-human-settlements-2011-cities-and-climate-change (accessed January 1, 2021).

United Nations-Habitat (UN-Habitat) (2016a) *New Urban Agenda* (Habitat III, Quito). United Nations Conference on Housing and Sustainable Urban Development. https://uploads.habitat3.org/hb3/NUA-English.pdf (accessed January 1, 2021).

United Nations-Habitat (UN-Habitat) (2016b) *UN-Habitat III: Adopted Draft of the New Urban Agenda*. Quito, Ecuador.

Wang, R., Ren, C., Xu, Y., Lau, K.K.-L. and Shi, Y. (2018) 'Mapping the local climate zones of urban areas by GIS-based and WUDAPT methods: A case study of Hong Kong', *Urban Climate*, 24: 567–76. doi.org/10.1016/J.UCLIM.2017.10.001

World Health Organization (WHO) (2010) *Urban Planning, Environment and Health: From Evidence to Policy Action* (meeting report). Geneva: WHO Regional Office for Europe. Available from https://www.euro.who.int/en/health-topics/environment-and-health/Housing-and-health/publications/2010/urban-planning,-environment-and-health-from-evidence-to-policy-action (accessed January 1, 2021).

World Health Organization (WHO) and UN-Habitat (2016) Global Report on 'Urban Health: Equitable, Healthier Cities for Sustainable Development'. Geneva: WHO. Available from https://www.who.int/gender-equity-rights/knowledge/global-report-on-urban-health/en/ (accessed January 1, 2021).

World Meteorological Organization (WMO) (2003) *Our Future Climate* (WMO-No. 952). Geneva: WMO. https://library.wmo.int/doc_num.php?explnum_id=4493 (accessed January 1, 2021).

World Meteorological Organization (WMO) (2016) *Integrated Weather, Climate, Hydrology and Related Environment Services for Sustainable Cities*. United Nations Conference on Housing and Sustainable Urban Development (UN-Habitat III), Quito, Ecuador. https://library.wmo.int/doc_num.php?explnum_id=3075 (accessed January 1, 2021).

Xu, J. (2012) 'Analysis of the Guangzhou urban heat island effect', *Robotics and Applications (ISRA)*, 2012. IEEE Symposium. pp. 555–7.

Xu, Y., Ren, C., Ma, P., Ho, J., Wang, W., Lau, K.K.-L., Lin, H. and Ng, E. (2017) 'Urban morphology detection and computation for urban climate research', *Landscape and Urban Planning*, 167: 212–24.

8

DATA ANALYTICS AND MODELLING ACCESSIBILITY CHANGE OF HIGH-SPEED RAIL NETWORK DEVELOPMENT: A DOOR-TO-DOOR APPROACH

Lei Wang

Introduction

Since the first commercial high-speed rail (HSR) line was opened in Japan in 1964, HSR has been introduced differently in many counties (Givoni, 2006). In general, two modes of HSR services have been developed globally: the maglev train and the wheel-track train. Most countries have adopted the mode of wheel-track, but with different approaches to integrating with the existing conventional rail (CR) network (Campos and de Rus, 2009). The first option is to build an independent line dedicated to passenger services, which is not compatible with CR lines. The second type is backward compatible with CR lines. The third option is to upgrade CR lines to operate much cheaper HSR services at the speed of 200–250 km/h. Two issues are critically important in HSR planning and development and have significant impacts on regional accessibility: (1) the number and location of stations, and (2) the integration of HSR station with the rest of the transport network (Banister and Givoni, 2013). Having been developed for decades, CR stations are usually incorporated in urban areas. Thus, HSR stations upgraded from CR stations are easy to access by most citizens. The location of newly built HSR stations is much more complicated because the construction costs in urban areas are very high and HSR lines require relatively straight and flat trajectories (Wang, 2018). Therefore, most HSR stations were built in suburban areas; examples include Spain, Taiwan and Mainland China. However, how the HSR network has changed the existing uneven spatial accessibility of different cities and produced winners and losers are contentious issues.

Various measurements have been developed to evaluate the accessibility impacts of the planning and construction of HSR projects. Gutiérrez (2001) examined the effects of the Madrid-Barcelona-French border HSR using three indicators: weighted average travel time, economic potential, and daily accessibility. Kim and Sultana (2015) formulated a systemised measurement to evaluate accessibility changes in the Seoul metropolitan area from HSR development in South Korea. Levinson (2012) predicted the potential accessibility impacts of HSR planning in the United States. Substantial literature and policy documents on accessibility calculation have emphasised the shorter inter-city rail travel time between stations from HSR development (Chandra and Vadali, 2014; Jiao et al., 2014; Kim and Sultana, 2015). On the other hand, some authors have highlighted the effects on intra-city travel time due to changes in the location of rail stations (Bellet, 2016; Givoni and Banister, 2012; Guirao, 2013). From the perspective of HSR users, some studies were concerned more about a journey's total travel time than the time of rail travel. Martínez and Givoni (2012) argued that accessibility assessment should not just focus on urban areas and stations; a wider geographic area must be considered in the multi-transport modes environment. Access to rail stations within a city's travel demand varies spatially, and HSR stations must be integrated with an intra-city transport network (Givoni and Banister, 2012; Wang et al., 2013). Even within the rail system, the integration between HSR and CR networks should be considered in regional accessibility evaluation (Garmendia et al., 2012). Urban networks with different transfer modes must be incorporated as feeders to HSR in accessibility assessments (Ortega et al., 2012; Zhao and Li, 2017).

China has been keen on developing HSR networks for its vast territory, and high population density, as well as to tackle the capacity problem of CR networks (Chen, 2012; Wu et al., 2014). The country has established the largest HSR network worldwide, most of which has been newly constructed in the last decade. By the end of 2016, China had an HSR network of 22,000 km that was suitable for trains running at speeds of over 200 km/h. The length accounted for more than 60% of the global length of HSR (UIC, 2018). In eastern megaregions, the HSR network has expanded so rapidly that it has almost caught up with the CR network for passenger transport (Ollivier et al., 2014). Rather than upgrading CR stations to be compatible with the HSR network, many cities in China have chosen to build brand new HSR stations in suburban areas, which increases the pressure on intra-city accessibility to HSR services (Diao et al., 2016). Perl and Goetz (2015) regarded HSR development in China as a distinct comprehensive model along with two other global models: exclusive corridor (e.g. Japan and the UK) and hybrid models (e.g. Germany and France). HSR development is assumed to have a complex impact on regional accessibility and spatial restructuring because of massive HSR network expansions and China's different development and station and line location strategy (Cao et al., 2013; Chen, 2012; Chen et al., 2016).

Research methods

Study area

Located in eastern coastal China (Figure 8.1), the Yangtze River Delta (YRD) has long been a policy focus of China's central government. In 2016, the YRD had a population of 222.04 million and produced US$2622.03 billion GDP. After several rounds of regional planning, the YRD territory has expanded from 15 member prefectures in three provinces in the late 1990s to 25 prefectures in four provinces in the last regional plan of 2016. Considering the continuous expansion of the YRD and China's railway administrative territory, the focus of this chapter is the so-called 'expanded YRD' encompassing the four province units of Shanghai, Jiangsu, Zhejiang, and Anhui. The megaregion includes four more developed urban areas (Shanghai, Nanjing, Hangzhou and Hefei) as well as the peripheral areas of northern Jiangsu province, south-western Zhejiang province, and most of Anhui province.

The YRD's railway system is administrated by the Shanghai Railway Bureau, one of the largest among the 18 local branches of China Railway. Both CR and HSR networks are well developed. By April 2017, the length of the CR passenger network was approximately 7,000 km and the length of the HSR network was over 3,000 km. There are 192 passenger railway stations in 122 county-level units, of which 122 stations operate HSR services for 99 county-level units.

Multi-scalar spatial analysis is used to conduct regional accessibility assessment. The basic spatial unit is the 100 m × 100 m grid and the YRD is covered by

34.81 million grids. Another spatial analytical unit is the county-level units. The grids are used as the basic building blocks for aggregating measures of accessibility change at different administration units. The county-level unit is the lowest administrative level for rail station construction in China. Additionally, regional and sub-regional level analysis is performed to evaluate the spatial variations of accessibility in the YRD.

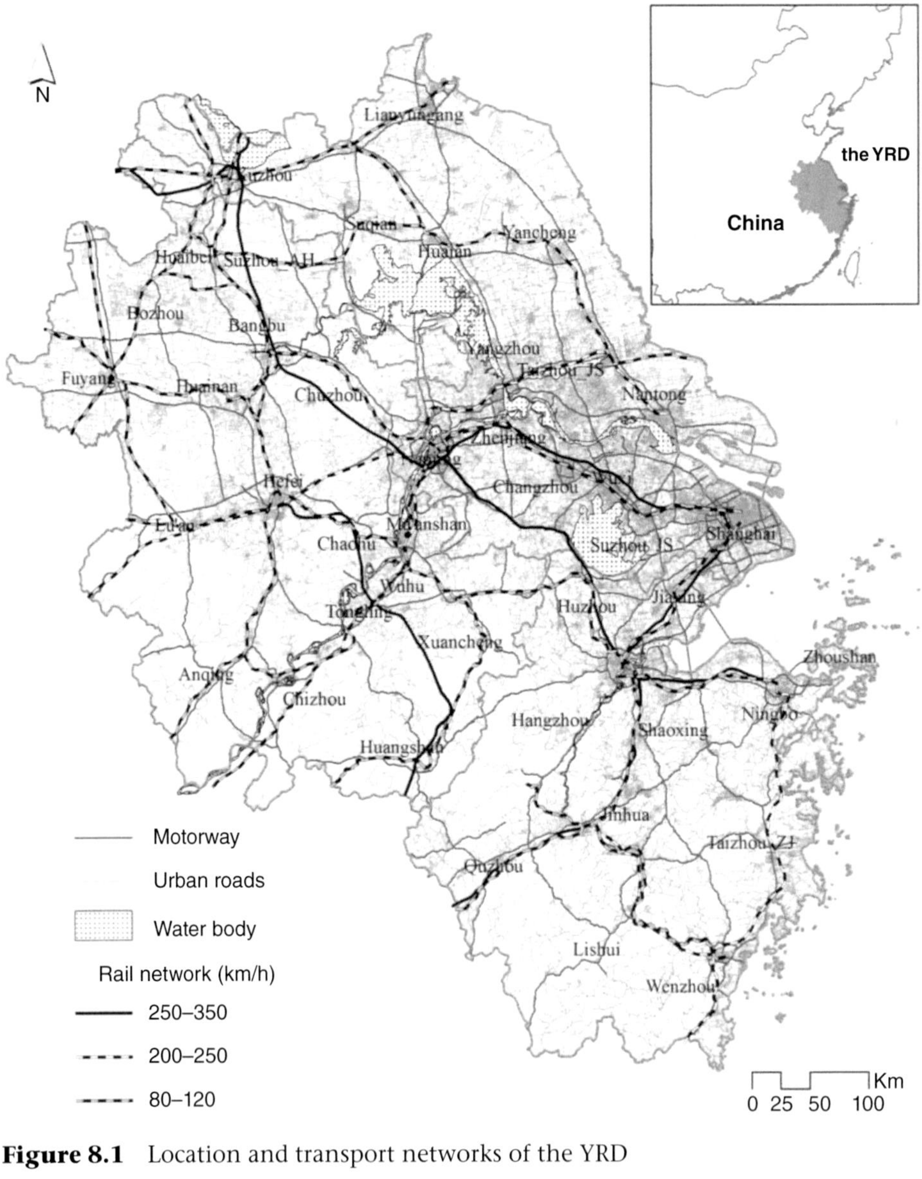

Figure 8.1 Location and transport networks of the YRD

Research steps of modelling winner and loser cities

The winner and loser cities are identified via a bottom-up approach by considering each city's relative accessibility gain or loss in relation to the average regional change. Previous studies typically used population as an endogenous factor of accessibility and employed it to weight the accessibility index and then compared accessibility values before and after HSR operation (Gutiérrez, 2001; Jiao et al., 2014). However, in this chapter, we regard residents themselves as losers or winners in HSR development. A city's performance depends on the aggregated situations of its residents, which are spatially unevenly distributed. The winner and loser residents are recognised by their travel time saving, which might be higher or lower than the average improvement of HSR development in the YRD. The travel time saving is calculated by comparing two scenarios of HSR-led regional accessibility and CR-led regional accessibility. The CR-led regional accessibility patterns are obtained by assuming the absence of HSR service, while HSR-led regional accessibility patterns are calculated on the basis of real operation of both HSR and CR services in the YRD. Figure 8.2 outlines the methodological steps and components of the research.

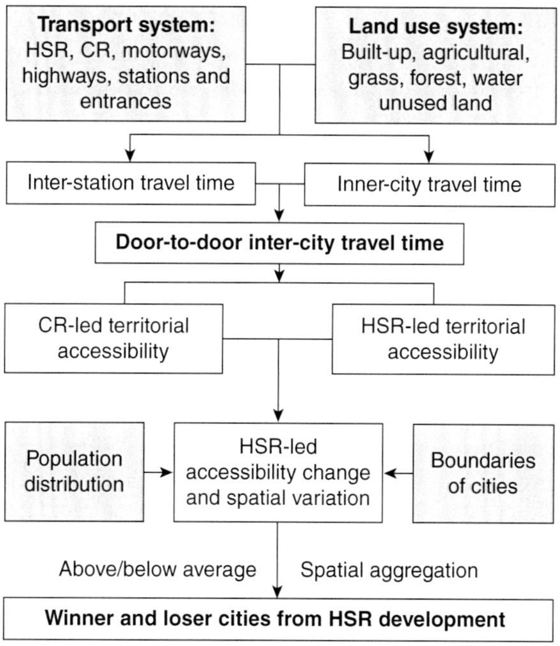

Figure 8.2 Research components and research steps

Key accessibility measures and data sources

Based on previous studies (Salonen and Toivonen, 2013; Wang et al., 2016), a door-to-door approach is employed to calculate the inter-city travel time. The total travel time from an origin city *i* to a destination city *j* includes three parts: (1) intra-city travel time to station before departure (T_1); (2) inter-city travel between two stations (T_2); and (3) intra-city travel after arrival from the rail station to the destination (T_3).

For modelling a journey's travel time in a Geographic Information System (GIS) environment (Monzón et al., 2013), the cost-distance analysis and network analysis are the two major methods, and they are operated with a raster dataset and a vector dataset respectively. By integrating the two methods, a hybrid approach is used to calculate regional accessibility to overcome the crossover problem in the road network for cost distance analysis and the lack of spatial resolution for network analysis. Specifically, the intra-departure city travel time (T_1) and intra-arrival city travel time (T_3) are calculated by a cost-distance analysis of the urban road system, whereas the inter-station travel time (T_2) is calculated through a network analysis of the rail network.

For the cost-distance analysis of intra-city travel time, we establish the following calculation procedure: (1) building layers of travel cost raster on different land-use types, transport networks, and infrastructures and calculating the minimum travel cost of all grids for all possible travel modes; (2) given railways and motorways are semi-closed systems, a two-sided buffer of barriers with low travel speeds is defined to ensure that the journey starts from rail stations and motorway entrances, respectively. A transfer time of 30 minutes is used and buffered around rail stations when a trip involves a transfer to rail services from road networks and a transfer time of 5 minutes for motorway entrances when a journey transfers from ordinary roads; and (3) each grid is allocated to a station catchment that is zoned by minimum travel cost; this resembles the reality as people tend to choose a rail service that is close to their residence.

For the network analysis of inter-city travel time between stations, the travel speeds of different train services in the YRD are used. We use the average travel speed of different types of train services to model rail travel time (Table 8.1). The speeds of other transport modes and on different land-use types are also regulated according to local design specification for road infrastructures and previous literature (Wang and Duan, 2018). Additionally, a one-time transfer of 30 mins is added if there is no direct train service between any two cities, and the information is obtained from official train timetables. Similarly, the principle of minimum travel time is used in the network analysis.

We measure regional accessibility by taking the total travel time to a group of destinations in the region. Similar to many studies (Monzón et al., 2013; Wang et al., 2013), we choose a group of major cities in the YRD to represent the destination. There are 42 cities at the prefecture-level and above[1] that are spatially distributed in the YRD. Finally, we define the physical distance to a city as the road and rail distance to its city centre. Thus, regional territorial accessibility of a grid *i* in the YRD is defined as follows:

[1]Administration at the level of prefecture or above in China indicates a city-region because a group of small cities is administered by a central city. The central city is usually the political and economic centre of the administration.

$$RT_i = min \sum_{j=1}^{n} T_{ij} \quad (8.1)$$

The purpose of transportation is to satisfy a demand for mobility, as transportation can only exist if people have to travel (Rodrigue et al., 2016). We further use population size to reflect the transport demand of a place. The geographic datasets of the transport network and facilities, land-use, and population have been collected from different sources. The transport data like types of road, railway, stations, and motorway entrances mainly come from the 2016 electronic atlas of the local and national transport departments in 2016. Still, we updated the rail system data based on Google Earth, as of April 2017, to reflect the most recent HSR development in the YRD. The administration boundaries and city centres are collected from national administration maps. Access to most recent high-resolution data on land-use and population is rather difficult. Only datasets from 2015 are available for both categories. The datasets for land-use and population distribution come from the Global Land Cover Datasets (GlobeLand30)[2] and WorldPop[3] spatiotemporal change dataset, respectively. Details of data sources and treatment are provided in Table 8.2. The Global Land Cover Datasets are also used to monitor land-use change and climate change in Chapter 7 of this book. In this chapter, land-use data are resampled into 100 m spatial resolution and reclassified into six types, as shown in Table 8.1.

Table 8.1 Modelling travel speeds of different transport modes and land-use types

Transport modes	Speeds (km/h)	Land-use types	Speeds (km/h)
HSR train I	300	Built-up area	20
HSR train II	250	Agricultural land	15
CR train	120	Grassland	10
Motorway	110	Forest land	2
National highway	80	Waterbody	5
Provincial highway	70	Unused land	5
City and country highway	40		
Streets	30		
Rail buffer of barrier	0.15		
Motorway buffer barrier	0.1		

[2]http://www.globallandcover.com/home_en.html (accessed January 4, 2021).

[3]https://www.worldpop.org (accessed January 4, 2021).

Table 8.2 Data treatment, description and sources

Data item	Data treatment and description	Data source
Rail networks and stations	The dataset includes CR and HSR networks and their stations that are in operation in 2017. HSR network is categorised into 2 types. Type 1 and 2 are operated at speeds around 300 km/h and 200 km/h respectively.	Extracted the rough data from OpenStreetMap, further cleaned and revised with the help of Google Earth (https://www.google.com/earth/) and Baidu Map. The dataset is free to download and manipulate.
Motorway network and entrances	High-resolution data. The direction of the vector data needed to be removed in the accessibility modelling. The entrances and their buffering zones needed to be defined.	Extracted from China Electronic Atlas in 2016 and further revised and topology-built with the help of Baidu Map. The dataset is free to download and manipulate.
Intra-city transport networks	The urban road network was categorised into levels of national, provincial, municipal and county with high spatial resolution. Each level has different speed limits and technical regulations. The data quality is varied across different cities and needed to be verified intensively.	Extracted the rough data from OpenStreetMap and further cleaned with the help of Baidu Map and national and provincial electronic atlases. The dataset is free to download and manipulate.
GlobeLand30 (Land use data)	The dataset is stored in raster format. The resolution is 100*100m grid. There are six land use types including agricultural land, forest land, grassland, water body, built-up land and unused land.	The data were purchased from GlobeLand30. The data were correlated with remote sensing images from Landsat Thematic Mapper. The most updated data are from 2015.
WorldPop (Population data)	The dataset has a high spatial resolution (100*100m grid). It disaggregated population census data at the administrative units to grid level with the help of satellite images and other sources.	The data are free to download from the WorldPop website. The most updated data are from 2015.
Administration data (city centres and boundaries)	The boundaries of administrative units of provinces, prefectures and counties. The city centre is regarded as the place where the municipal government is located. We use the 2010 data due to frequent changes to boundaries and city centres in the past decades in China.	The data of administrative boundaries were purchased from National Geomatics Centre of China.The city centres were crawled from Baidu POI services.

Sources: Baidu Map (https://map.baidu.com/), GlobeLand30 (http://www.globallandcover.com), Google Earth (https://www.google.com/earth/), Landsat Thematic Mapper (https://landsat.gsfc.nasa.gov/landsat-4-5/tm), National Geomatics Centre of China (http://ngcc.cn/), OpenStreetMap (www.openstreetmap.org)

Results

Absolute and relative changes in regional accessibility

Regional accessibility of total travel time to major cities in the YRD is measured according to the formula 8.1. Natural breaks of travel time and its savings from HSR development were used to classify city groups. For the accessibility modelling for CR network, the highest territorial accessibility with total travel time ranging from 124 to 163 hours is mainly clustered in the geographic centre of the region (e.g. cities of Nanjing, Zhenjiang, and Ma'anshan). The areas in various corridors radiating from the geographic centre are highlighted, as in the case of Changzhou, Wuxi, and Huzhou in the east and Bangbu, Huainan, Hefei, and Tongling in the west. The development of HSR network improved territorial accessibility significantly. The highest territorial accessibility (88–133h) increased by one-third and expanded dramatically to account for 18.4% of YRD territory. Higher territorial accessibility values are mainly located in the areas around HSR stations. Besides reappearing accessibility centres obtained in the CR network, some enclaves with higher values also emerge in the periphery areas, such as Xuzhou, Liu'an, Huangshan, Quzhou, and Jinhua. The pattern shows that the HSR network helps geographically peripheral cities (e.g. Xuzhou, Ningbo, and Shanghai) to overcome their barriers of long rail distance. Correspondingly, the lowest regional accessibility (with only 1.3% of the territory) shrinks drastically and is mainly located in economically backward areas, mountain areas, and water bodies with few residents, such as the northern Jiangsu province and the southwestern Zhejiang province. The accessibility level increased throughout the territory, and the average improvement was 23.4%. The time saving of door-to-door travel is significantly greater than that of the inter-city rail travel (51.2%).

Territorial accessibility change from the HSR network varies across the region, the stations, and their surrounding areas. It is not surprising that high absolute accessibility change is mainly located in the YRD's periphery areas because their regional accessibility is determined by relatively large rail distances (Figure 8.3a). Besides, the outstanding accessibility gain for the coastal area in southern Zhejiang province is also due to development of the Ningbo-Taizhou-Wenzhou HSR line. We further mapped relative change of travel time from CR to HSR network to remove the scale effect for a fairer comparison. As shown in Figure 8.3b, the areas with the greatest improvement (33.8–52.7%) are scattered across the region and are found in two types of areas. The first type is mainly found in areas where there were no rail stations before the HSR development. For example, cities along the Ningbo-Taizhou HSR and the Nanjing-Hangzhou HSR belong to this type. The second type is some HSR cities located in the peripheral areas where HSR services help to overcome their location disadvantage. Besides, comparing with CR lines, HSR lines are much straighter, which reduces rail distance between cities.

However, according to the statistics on the distribution of relative travel time saving, over 70% of the territory's change value is below 30%. It is interesting to find that some HSR cities belong to the group with the lowest territorial accessibility change, such as Caohu, Suzhou-AH, and Jiaxing. These HSR cities show a change of less than 20% due to the poor connections from HSR stations, which means that the cities benefit less than they were supposed to from the introduction of the HSR network. In Suzhou-AH in Anhui province, for example, the city's HSR station is 31 km away from the city centre. The HSR station was built on farmland and poorly connected with nearby cities and towns, which prevents it from creating the benefits associated with inter-city travel via HSR service. The journey by car from the city centre takes less than 5 minutes to reach the CR station, but 42 minutes to reach the HSR station. Similarly, for Jiangxing city in Zhejiang province, the journey from the city centre takes only 10 minutes by car and 30 minutes by bus to the CR station respectively, while the figures increase to 20 and 60 mins for travel to the HSR station. The result suggests that intra-city transportation to the HSR station is critical when considering territorial accessibility improvement of the region.

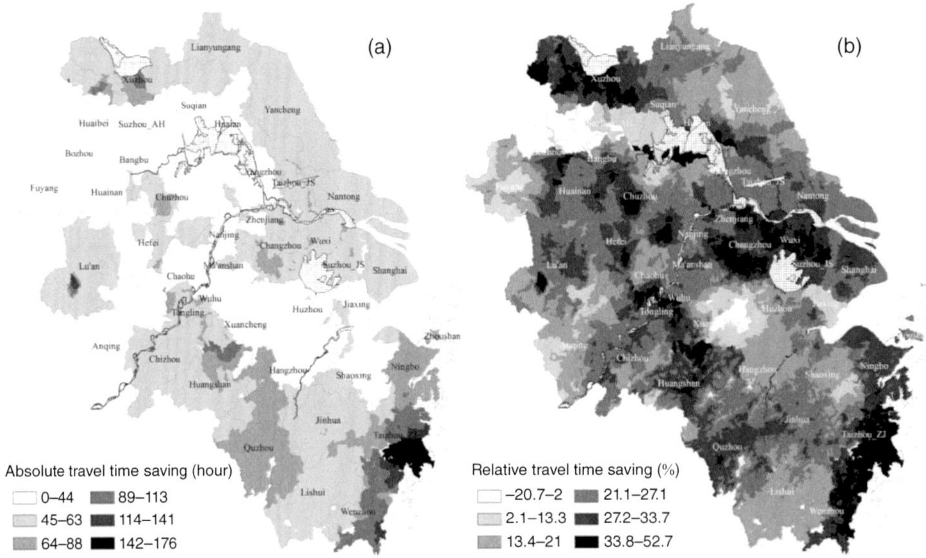

Figure 8.3 Patterns of travel time saving from HSR compared to CR network

Winner and loser cities on regional accessibility improvement

Based on the pattern of changes in regional accessibility, we explore the winner and loser cities by aggregating the respective winner and loser population of accessibility improvement at the grid level. There are four steps: (1) locating the winner and loser grids based on their regional territorial accessibility change above and below the average percentage of improvement (Figure 8.4a); (2) weighting the winner and loser grids by their population size (Figure 8.4b) to establish winner and loser populations at the grid level; (3) aggregating net winner populations through city boundaries; and (4) defining winner cities with a threshold of net winner population of at least half of the total population (Figure 8.4c).

Figure 8.4c shows the distribution of net winner and loser population across cities in the YRD as a result of the HSR network development. The winner cities from HSR network development are mainly located in places that have developed HSR and CR networks (e.g. the mid-YRD) and in the geographically peripheral areas with HSR stations (e.g. coastal Zhejiang and north-western Jiangsu). In contrast, most losers are those cities with no HSR service within their jurisdictions or in their vicinity. However, there are some unusual HSR cities that are losers, such as Suzhou-AH. Our evaluation shows that the existence of an HSR station is not the only factor determining whether a city benefits from HSR network development, the location of the HSR station and its accessibility to most people within the city are also important. The YRD exhibits seven types of winner and loser cities with different spatial patterns for territorial accessibility improvement, HSR station location and population distribution.

Winners of Type I are mainly the megacities, which are also rail hubs with several CR and HSR stations covering most of the population. This type is exampled by Shanghai, Nanjing, Hefei and Hangzhou, with several rail stations that can be easily accessed by most of the city's population, and some HSR stations that were upgraded from CR stations close to the city centre. Winners of Type II have HSR stations upgraded from CR stations and close to the city centre. They are exampled by Anqing in Anhui province. Most cities in the upgraded Nanjing-Nantong Line and the cities of Quzhou and Jinhua with upgraded CR stations for the new HSR line belong to this type. Winners of Type III are represented by Taizhou-ZJ, which had no rail services before the introduction of HSR. HSR stations are typically built in suburban areas to save construction costs, but this significantly diminishes the city's rail accessibility. Winners of type IV are small cities such as Taicang, Changshu, Jingjiang, and Pinghu, which benefit from the spill-over effect of neighbouring HSR services. This implies that it is not necessary for some small cities to build costly HSR stations, and it is often more efficient to build a relatively cheap motorway to improve its accessibility by connecting to nearby HSR stations.

Turning to the loser cities, Type V cities have an HSR station but far from the city centre. Examples of this type are Shaoxing in Zhejiang province and Suzhou and Huai'an in Anhui province. The result shows that an HSR station does not secure better regional accessibility improvements if the station is located far from the city centre and is less accessible for most people. Losers of type VI and VII are marginalised by the development of an HSR network because of their remote locations and underdeveloped economies. The relative distance of an HSR station from a CR station plays a vital role in deciding whether an HSR city is a winner or a loser. This is because CR stations have shaped urban development for decades in the YRD, and the surrounding areas either became the city centre or attracted a large population. Therefore, most CR stations in the YRD are accessible to the city. In contrast, many newly built HSR stations are in suburban areas and are less accessible. Additionally, with only a decade of development, the stations' surrounding areas are far less developed in some mid-sized and small HSR cities.

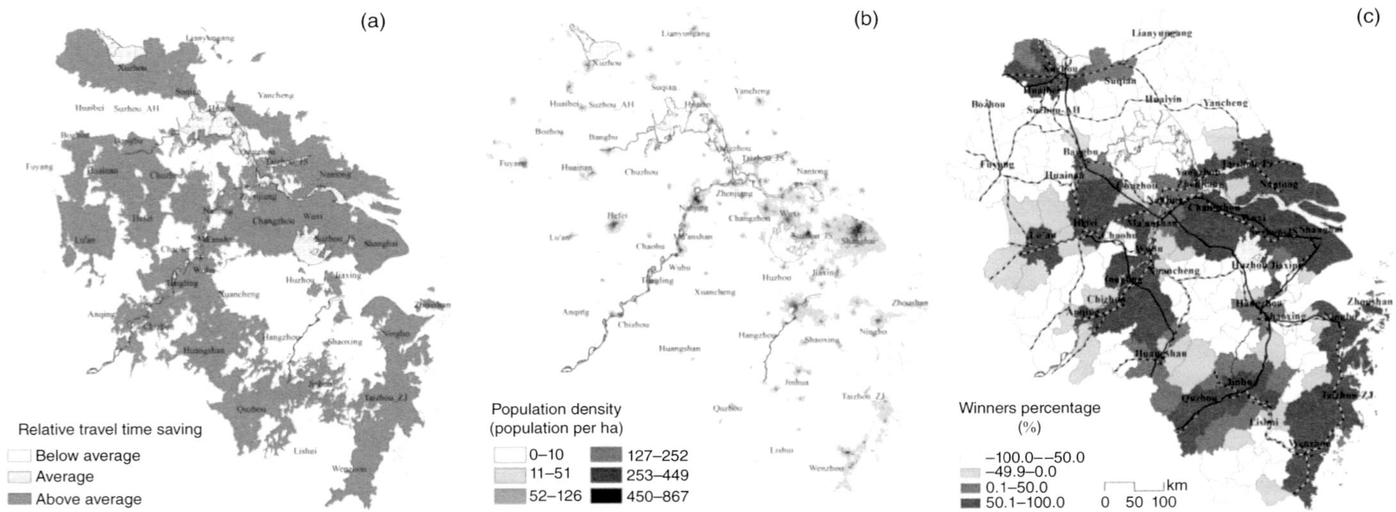

Figure 8.4 Average accessibility improvements, resident distribution and winner and loser cities in the YRD

Conclusion

The development of the HSR network has significantly improved regional accessibility and connectivity in the YRD. However, what is contentious is whether major infrastructural network investment has created an uneven landscape of winners and losers. Various studies have highlighted time compression in inter-city rail travel from HSR network development but have overlooked the change in the location of rail stations, which prolongs intra-city travel time and counteracts the role of HSR services. Based on the emerging high spatial resolution of multiple datasets of transport, land-use and population, this chapter develops a door-to-door accessibility measurement by integrating the methods of cost-distance analysis and network analysis. The hybrid approach can overcome the crossover problem in the road network for cost-distance analysis and the low spatial resolution in network analysis. This chapter also contributes to our understanding of regional accessibility improvement by incorporating the often ignored intra-city travel time to railways stations outside the HSR network.

Both CR and HSR networks generate a core-periphery structure and corridor effects of territorial accessibility in the YRD, but the HSR network exhibits more powerful spatial influences on spatial structure and creates many high accessibility enclaves around HSR stations in the periphery areas. The findings show that the geographic location, together with the rail network extension determines the patterns of regional territorial accessibility. HSR development has compressed regional travel time significantly, but the improvements vary across different sub-regions, cities, and stations. Some of the highest gains are in peripheral areas and main HSR corridors, while some HSR cities gain territorial accessibility improvements that are below average. Intra-city accessibility of HSR stations also plays an important role in regional accessibility improvements. The insignificant change in the spatial unevenness of territorial accessibility from the HSR network to the CR network further suggests the limited role of inter-station accessibility improvements in inter-city accessibility. However, the HSR network decreases the spatial unevenness of overall population accessibility significantly. Particularly, the development of intermediate stations along the Nanjing-Hangzhou HSR and Ningbo-Wenzhou HSR had given residents in cities like Liyang, Yixing and Taizhou-ZJ access to rail services for the first time. These cities are emerging developed cities with relatively high population density and advanced industries. Therefore, the newly built HSRs have improved rail accessibility for a greater population and thus impacted on the social equity of rail access in the YRD.

In contrast to the experience in Europe where most HSR stations are CR stations upgraded for accommodating HSR trains (Banister and Givoni, 2013), most HSR stations in the YRD are newly built and located at the edge of urban areas. The development strategy of China has complicated the intra-city connectivity and accessibility to HSR stations. Consequently, the existence of an HSR station does not necessarily mean that a city will be a winner in regional accessibility, as the relative location of an HSR station from the city centre and its proximity to major population centres also plays an important role. The results suggest that poor access to the HSR station is a common issue for many cities and more integrated development is needed for future HSR planning and development.

As argued by Wong in Chapter 3 of this book, big open datasets provide the possibilities and new resources for scholars to understand the spatial dynamics of urban regional development, but they should also be processed, evaluated and analysed to unpack the fog of information sources. This argument is partly evidenced in our case study because various data sources are used in this chapter, but their accuracy and reliability is unknown and probably varies spatially. We use OSM in this study, but the accuracy of this data source varies significantly across urban and rural areas and developed and developing countries. Fortunately, the YRD is one of the most developed megaregions in China and the OSM data are complemented by a land-use dataset to model territorial accessibility. The situation of the population dataset is similar. The population surface of WorldPop was disaggregated from various countries' census data by making reference to a number of geospatial datasets such as satellite imagery and cell phone data (Tatem, 2017). In this chapter, only the overall accuracy rate has been examined by sampling locations in the online map platform. This uneven data provision of different places and different sources becomes a critical issue for spatial comparisons, which needs to be carefully dealt with in future urban and regional studies.

Some limitations of this method of modelling accessibility should be mentioned. Firstly, this chapter does not take the availability of vehicle supply and waiting time for traffic lights and traffic jams into account in intra-city travel time calculations. Although our sample locations illustrate that the model results exaggerate the accessibility of railway stations by car by approximately 10%, the gap may be larger in rural areas as road conditions are relatively poor. The accessibility gap between public and private transport to HSR stations should also be evaluated for future spatial equity analysis. The issue is particularly important in the context of developing countries where most people use public transport as their first choice of travelling. Secondly, there is a need to reconsider the least travel time principle for zoning railway catchment. Practically, the factors affecting travellers' station selection also include the frequency and cost of train service. In future studies, time-series accessibility could be acquired from online map queries. As we know, when considering inner-city accessibility, travel time differences between rush hours and off-peak hours are significant and should be examined in future accessibility analysis.

Acknowledgements

This research is funded by Hallsworth Research Fellowship Fund at the University of Manchester and National Science Foundation of China (41601169). The author would like to thank Cecilia Wong for her insightful suggestions and comments.

References

Banister, D. and Givoni, M. (2013) 'High-speed rail in the EU27: Trends, time, accessibility and principles', *Built Environment*, 39: 324–38.

Bellet, C. (2016) 'Peripheral high-speed rail stations in Spain', *The Open Transportation Journal*, 10: 45–56.

Campos, J. and de Rus, G. (2009) 'Some stylized facts about high-speed rail: A review of HSR experiences around the world', *Transport Policy*, 16: 19–28.

Cao, J., Liu, X.C., Wang, Y. and Li, Q. (2013) 'Accessibility impacts of China's high-speed rail network', *Journal of Transport Geography*, 28: 12–21.

Chandra, S. and Vadali, S. (2014) 'Evaluating accessibility impacts of the proposed America 2050 high-speed rail corridor for the Appalachian Region', *Journal of Transport Geography*, 37: 28–46.

Chen, C.-L. (2012) 'Reshaping Chinese space-economy through high-speed trains: Opportunities and challenges', *Journal of Transport Geography*, 22: 312–16.

Chen, Z., Xue, J., Rose, A.Z. and Haynes, K.E. (2016) 'The impact of high-speed rail investment on economic and environmental change in China: A dynamic CGE analysis', *Transportation Research Part A: Policy and Practice*, 92: 232–45.

Diao, M., Zhu, Y. and Zhu, J. (2016) 'Intra-city access to inter-city transport nodes: The implications of high-speed-rail station locations for the urban development of Chinese cities', *Urban Studies*, 54 (10): 2249–67.

Garmendia, M., Ribalaygua, C. and Ureña, C.M. (2012) 'High speed rail: Implication for cities', *Cities*, 29 (Supplement 2): S26–S31.

Givoni, M. (2006) 'Development and impact of the modern high-speed train: A review', *Transport Reviews*, 26: 593–611.

Givoni, M. and Banister, D. (2012) 'Speed: The less important element of the High-Speed Train', *Journal of Transport Geography*, 22: 306–7.

Guirao, B. (2013) 'Spain: Highs and lows of 20 years of HSR operation', *Journal of Transport Geography*, 31: 201–6.

Gutiérrez, J. (2001) 'Location, economic potential and daily accessibility: An analysis of the accessibility impact of the high-speed line Madrid–Barcelona–French border', *Journal of Transport Geography*, 9: 229–42.

Jiao, J., Wang, J., Jin, F. and Dunford, M. (2014) 'Impacts on accessibility of China's present and future HSR network', *Journal of Transport Geography*, 40: 123–32.

Kim, H. and Sultana, S. (2015) 'The impacts of high-speed rail extensions on accessibility and spatial equity changes in South Korea from 2004 to 2018', *Journal of Transport Geography*, 45: 48–61.

Levinson, D.M. (2012) 'Accessibility impacts of high-speed rail', *Journal of Transport Geography*, 22: 288–91.

Martínez, H.S. and Givoni, M. (2012) 'The accessibility impact of a new High-Speed Rail line in the UK – a preliminary analysis of winners and losers', *Journal of Transport Geography*, 25: 105–14.

Monzón, A., Ortega, E. and López, E. (2013) 'Efficiency and spatial equity impacts of high-speed rail extensions in urban areas', *Cities*, 30: 18–30.

Ollivier, G., Bullock, R., Ying, J. and Zhou, N. (2014) High-speed railways in China: A look at traffic (China Transport Topics No. 11) Beijing: World Bank. Available from https://openknowledge.worldbank.org/handle/10986/25480 (accessed January 4, 2021).

Ortega, E., López, E. and Monzón, A. (2012) 'Territorial cohesion impacts of high-speed rail at different planning levels', *Journal of Transport Geography*, 24: 130–41.

Perl, A.D. and Goetz, A.R. (2015) 'Corridors, hybrids and networks: Three global development strategies for high speed rail', *Journal of Transport Geography*, 42: 134–44.

Rodrigue, J.-P., Comtois, C. and Slack, B. (2016) *The Geography of Transport Systems* (4th edition). London: Taylor & Francis.

Salonen, M. and Toivonen, T. (2013) 'Modelling travel time in urban networks: Comparable measures for private car and public transport', *Journal of Transport Geography*, 31: 143–53.

Tatem, A.J. (2017) 'WorldPop, open data for spatial demography', *Nature: Scientific Data*, 4: article #170004.

UIC (International Union of Railways) (2018) *High-speed Rail. Fast Track to Sustainable Mobility*. Paris: International Union of Railways. https://uic.org/IMG/pdf/uic_high_speed_2018_ph08_web.pdf (accessed January 4, 2021).

Wang, J., Xu, J. and He, J. (2013) 'Spatial impacts of high-speed railways in China: A total-travel-time approach', *Environment and Planning A*, 45: 2261–80.

Wang, L. (2018) 'High-speed rail services development and regional accessibility restructuring: The case of the Yangtze River Delta, China', *Transport Policy*, 72: 34–44.

Wang, L. and Duan, X. (2018) 'High-speed rail network development and winner and loser cities in megaregions: The case study of Yangtze River Delta, China', *Cities*, 83: 71–82.

Wang, L., Liu, Y., Sun, C. and Liu, Y. (2016) 'Accessibility impact of the present and future high-speed rail network: A case study of Jiangsu Province, China', *Journal of Transport Geography*, 54: 161–72.

Wu, J., Nash, C. and Wang, D. (2014) 'Is high speed rail an appropriate solution to China's rail capacity problems?', *Journal of Transport Geography*, 40: 100–11.

Zhao, P. and Li, S. (2017) 'Bicycle-metro integration in a growing city: The determinants of cycling as a transfer mode in metro station areas in Beijing', *Transportation Research Part A: Policy and Practice*, 99: 46–60.

9

UNDERSTANDING MIGRANT–LOCAL NEIGHBOURLY RELATIONS IN SHANGHAI: EMPIRICAL URBAN RESEARCH IN A DATA-SPARSE SETTING

Zheng Wang

Introduction

In recent years, undoubtedly new forms of data have helped to better understand the relationship between the built environment and important challenges such as climate change (Ren et al., Chapter 7 in this volume) and crowdsourced geographic information (Yudono, Chapter 4 in this volume). Likewise, against a background of growing migrant residents in cities, improving intergroup relations has also become an important means to maintain social cohesion and help migrants to socially integrate into the city. However, unlike for instance climate change, which can be measured through tangible indicators such air pollution or CO_2 emission, it is often harder to measure the social aspects of cities, which are based on intangible concepts such as interpersonal trust or tolerance. Data scarcity is often a major challenge when researching social trends and problems. In China, whilst there is an abundance of data concerned with social trends, such data is often inaccessible to the public and particularly to international research institutions and academics. Under such circumstances, questionnaire surveys have become an important means to gauge the social trends and challenges that contemporary cities face.

In China, the social integration of rural migrants, who moved from China's rural areas to Chinese cities, is one of the greatest challenges for Chinese authorities. There are more than 245 million rural migrants living in Chinese cities (NBS, 2017) and whilst all rural migrants are also Han Chinese, their rural registration status means that they are unable to access welfare entitlements in cities (e.g. public housing). The Chinese government and many research institutions have carried out many large-scale household surveys to better understand the challenges faced by rural migrants. However, the reason why I argue that China in many ways is still a 'data scarce' context is because such data sets are often not publicly accessible. A further downside is that existing survey data sets do not address the vast number of social issues and challenges that rural migrants face. Under such circumstances, carrying out your own questionnaire survey has become one of the most effective ways to study emerging social issues in urban China.

The purpose of this chapter is to explain how a household survey has helped to investigate a pressing social issue in Chinese cities, namely the intergroup relationship between rural migrants and native urban residents. Studies show that rural migrants are suffering from discrimination by native urban residents and feel socially excluded from the host society (Chen et al., 2011; Nielsen and Smyth, 2011). Poor intergroup relations with native residents characterised by stigmatisation and discrimination have also led to serious mental health problems for rural migrants (Chen et al., 2011; Liu et al., 2017a, 2017b; Nielsen and Smyth, 2011). However, so far there are few studies that have examined what affects the intergroup relations between migrants and local residents in urban China and how neighbourhood factors such as the share of migrant residents might affect this outcome. The purpose of this chapter is to investigate the drivers and inhibitors of intergroup social relations between migrant and native residents using Shanghai as the case study. The study uses neighbourly interactions such as visiting each other's homes and mutual support between rural migrants and urban residents as a proxy of the intergroup relationship between migrants and urban residents.

The chapter is structured as follows: the next section reviews the existing research on intergroup relations both internationally and in the case of urban China. This is followed by a description of the research method and the survey data and then the findings are reported. The final part discusses the context of the findings and how they can contribute to the understanding of intergroup relations.

The drivers of intergroup relations in urban China

The integration of internal migrants is a longstanding research topic that has been studied both historically e.g. during the industrial age (Moch, 2003), but also in contemporary fast developing countries such as in sub-Saharan Africa (Gugler, 2002), Malawi (Englund, 2002) and Turkey (Erman, 1998). Intergroup relations have mostly been research in a multi-ethnic context in order to better understand the social integration of immigrants and minority groups as well as their relationship with majority group members (Laurence, 2011; Laurence and Bentley, 2016; Letki, 2008; Putnam, 2007). Existing studies found that despite belonging to the same Han ethnicity, migrants who have come from more rural areas are especially exposed to unequal treatments by the state as well as discrimination by native urban residents (Nielsen and Smyth, 2011; Nielsen et al., 2006). Rural migrants are stigmatised with the image as lowly educated and poor individuals with a tendency to criminal behaviour (Chen et al., 2011; Li and Wu, 2013). Despite being from the same Han ethnicity, the social exclusion and marginalisation of rural migrants have resulted in two distinct social groups, namely migrants and locals. Research has shown that migrant and local residents in urban China differentiate strongly between a *us* (in-group) and *them* (out-group) and do not consider each other as belonging to the same social group (Nielsen et al., 2006; Wang et al., 2016; 2020).

To study the relationship between native and migrant residents, recent studies have focused on the local social interactions of rural migrants (Huang et al., 2018; Wang et al., 2016, 2017a, 2017b; Wu and Logan, 2016; Wu et al., 2019). In the Chinese context existing studies found that the hukou (registered residency) status of an individual is a major determinant of neighbouring, where rural migrants tend to have more neighbourly interactions compared to native residents (Wu and Logan, 2016). Being excluded from any welfare entitlements, rural migrants are marginalised and have few social ties with natives (Liu et al., 2013) and are thus more reliant on localised forms of social ties. Stigmatisation, which increases social distance between locals and migrants, is often stated as the key reason why rural migrants struggle to acquire native acquaintances and friendship (Chen et al., 2011; Yue et al., 2013). Many native residents often associate migrants from rural areas with poor education as well as poverty and criminal behaviour (Chen et al., 2011; Li and Wu, 2013). From this perspective it is possible to state that the rural hukou is not only connected with institutional limitations but also with a negative stigma, which bears great similarity with for instance belonging to being Black in America (Alesina and La Ferrara, 2002). In this sense, the rural hukou can also be considered as a proxy indicator for the perceived social distance between migrants and locals.

In addition to hukou status, neighbouring studies in urban China also identified other individual level characteristics that are significantly related with the frequency of general neighbouring activities and their findings are in accordance with most neighbouring literature from the West (Buonfino and Hilder, 2006; Guest and Wierzbicki, 1999). For example Li et al. (2012: 250) found that in Guangzhou a resident's length of residency is positively associated with neighbourly interactions whilst education is a negative predictor. Similarly, Forrest and Yip (2007) also suggest that residents who have lived more than ten years in the locality tend to have more neighbourly relations. The reason is because their long residency has given them a stronger sense of belonging to the locality (Guest and Wierzbicki, 1999). Furthermore, Li et al. (2012) argue that higher socio-economic status in general would lead to less frequent neighbourly interactions. Hazelzet and Wissink's (2012) study further confirms that residents with a higher level socio-economic status tend to be less involved in the neighbourhood.

In addition to individual level determinants, research in urban China also indicates that neighbourhood factors are related to the neighbouring patterns of residents. Forrest and Yip (2007) found that the frequency of neighbourly interactions is high in older and more established neighbourhoods such as traditional courtyard homes and work-unit housing. In comparison, neighbouring activities decrease considerably in newly built commodity housing (Forrest and Yip, 2007). Residents in older neighbourhoods tend to interact more with their neighbours because they are more accustomed to a neighbourhood-based social life, which was omnipresent in the work-unit era of China (Hazelzet and Wissink, 2012). Additionally, residents living in older neighbourhoods also tend to be poorer (Wu et al., 2010) and therefore rely more on localised social networks. Comparatively, recent findings indicate that residents living in commodity housing are less keen on interacting with their neighbours but instead regard their neighbourhood as a source for privacy and security (Zhu et al., 2012), but also as a sign of social status (Pow, 2007).

In multi-ethnic societies the residential diversity of an area is often measured through the share of minority residents, which is a major determinant of intergroup neighbouring (Gundelach and Freitag, 2014; Ihlanfeldt and Scafidi, 2002; Stolle et al., 2008; Twigg et al., 2010). In regards to the effect of residential diversity, there are two main arguments which contradict each other. The *contact theory* coined by Allport (1954) argues that the more individuals are exposed to an environment with people from other social groups, the more likely it is that they will have positive intergroup relations (Allport, 1954; Pettigrew, 1998). In contrast, the *conflict theory* suggests that higher levels of ethnic diversity in an area can reduce intergroup neighbourly interactions (Guest et al., 2008; Laurence and Bentley, 2016; Putnam, 2007). In urban China there is no empirical evidence showing the impact of residential diversity, but it is clear that some neighbourhoods have higher numbers of migrant residents compared to others (Li and Wu, 2008; Shen and Xiao, 2020). Therefore, the presence of migrant residents will be considered as an important neighbourhood level determinant in this study. In addition, neighbourhood poverty is also considered an important determinant whereby again ethnic minorities living in deprived areas are especially isolated from neighbourly contacts with native residents (Laurence, 2011; Letki, 2008). The reason is because ethnic minorities are often considered as a threat by locals who fear that minority neighbours may compete with them for the limited resources of the neighbourhood (Laurence, 2011). In Chinese cities poverty has also been found to concentrate in certain neighbourhoods where rural migrants and redundant state workers make up its two major population groups (Wu et al., 2010). Given the unanimous argument that poorer areas tend to have less intergroup neighbourly interactions, the analysis of this chapter will also include the poverty rate of neighbourhoods as a key neighbourhood factor (Laurence, 2011; Letki, 2008; Putnam, 2007).

Based on the brief review above, this study hypothesises the following:

H1) Migrant residents are more likely to engage in intergroup neighbourly relations compared to local residents.

H2) Neighbourhood characteristics play a significant role in determining the frequency of intergroup neighbouring activities whereby residents living in areas with a higher migrant presence is positively associated with higher neighbouring frequencies.

H3) The poverty rate of an area does not have a negative impact on the neighbourly interactions between migrants and locals.

Research methodology

Survey and data

This research is based on a household survey carried out in Shanghai in August 2013. Shanghai was chosen as a case study due to its high share of migrant population whereby 40% of the 23 million population are migrants (NBS, 2017). Furthermore, Shanghai's migrant population is very diverse in terms of socio-economic status and places of origin (Li and Wu, 2008; Shen and Xiao, 2020), which is helpful for the purpose of this study. Moreover, Shanghai's problems of discrimination towards non-locals are representative of the problems faced by many other Chinese cities. The survey was conducted by a group of professionally trained surveyors on a city-wide scale. The survey followed a two-stage random sampling strategy in order to generate a representative sample of the Shanghai population.

The sampling area was at the residential committee (*juweihui*) level, which is lower than the subdistrict level (*jiedao*) and is naturally defined by building blocks and streets. For the first stage of the sampling all sub-districts (including both urban and rural) were separated into three strata for the sampling, namely areas within the inner ring road, areas between the inner and outer ring roads and finally areas outside of the outer ring road. Areas within the inner ring road can be considered as the urban core of Shanghai, whereas areas between the inner and middle ring road represent the inner suburb of Shanghai. Finally, areas outside of the outer ring road can be considered as the outer suburban parts of the city. The three areas are distinctly different in terms of housing price and housing types, whereby the inner city is more likely to have older housing whilst the suburban areas are more likely to have new high rise residential neighbourhoods. The number of sub-districts sampled in each stratum is proportional to the overall population size in each area based on the 2010 population census data of Shanghai. For all three stratums, sub-districts were sorted according to their per capita income and population density, which ensured the representativeness of the sample. The survey then used the method of probability proportionate to size (PPS) sampling to choose individual sub-districts. Following this method, we then chose 33 sub-districts (out of a total of 225 sub-districts) from 12 districts in Shanghai. The second stage involved the random selection of residential committees within the chosen subdistrict in order to ensure that all types neighbourhoods have an equal chance of being chosen. Overall the survey contains data from 35 neighbourhoods which are distributed fairly evenly across Shanghai in order to examine the neighbourly relations in different types of neighbourhoods such as neighbourhoods with lower levels of migrant residents and areas with high concentration of migrant population. Forty questionnaires were distributed in each neighbourhood. Figure 9.1 shows the approximate location of all 35 neighbourhoods where the survey was conducted. Households were randomly chosen for interview based on a random starting street number with a fixed interval. Our survey

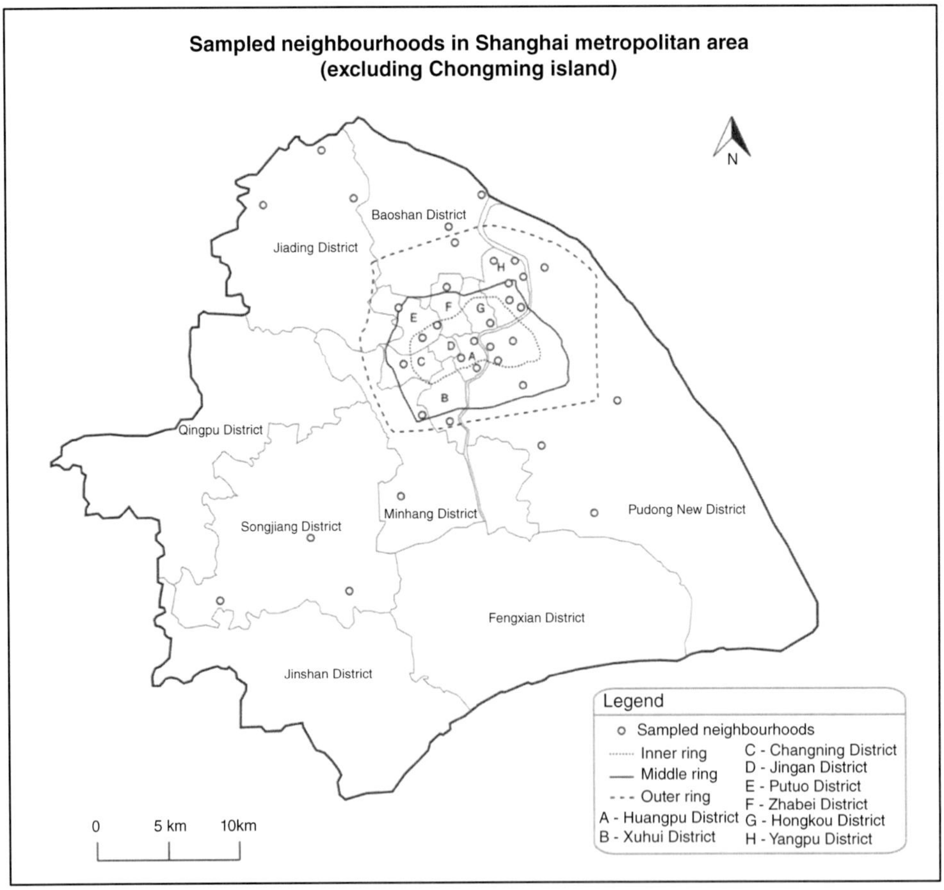

Figure 9.1 Sampled neighbourhoods in Shanghai

required the head of household to be interviewed and the survey yielded 1420 valid samples in total. The success rate for this survey was very high (95%) because members of the residential committee helped introducing the surveyors to selected households.

Amongst the sampled households 1046 are local urban hukou residents, 128 local rural hukou, 86 non-local urban hukou and 158 non-local rural hukou migrants. The reason why the migrant ratio is below the city's average, which was around 40% in 2013 (Shanghai Municipal People's Government, 2014), is because many migrant residents were unavailable for interviews due to their irregular work schedule and long working hours. In order to remediate this shortcoming after the initial survey another 100 migrant samples were added to avoid the possibility of leaving out any particular migrant groups. A comparison of the survey's sample and with official statistics published by Shanghai government, including the city's sixth population census and the Shanghai Statistical Yearbook 2014, reveals that there is no systematic lack of any particular respondent group (which can be seen in Table 9.1).

Table 9.1 Comparison of survey data and official statistics

	Survey data in 2013	Official statistics
Educational attainment of working age population		
Below elementary	0.64%	1.0%[a]
Elementary	5.26%	9.0%[a]
Junior secondary	33.59%	40.2%[a]
Senior secondary	21.67%	21.5%[a]
College or above	33.33%	28.3%[a]
Income per month	3548.53 Yuan	3654.25 Yuan[b]
Major occupational sectors	**Migrant sample only (N=243)**	**Migrants only**
Commercial and service industry staff	42.58%	31.5%[b]
Production and manufacturing	19.52%	28.7%[b]
Construction	8.10%	4.1%[b]
Transport and Logistics	4.29%	5.5%[b]

Source: [a] Shanghai sixth population census in 2010; [b] Shanghai Statistical Yearbook 2014

Definition of intergroup neighbourly relations

Neighbouring in this study is measured through three indicators, namely the frequency of visiting each other's homes, exchanging support amongst neighbours and greeting each other. We chose these three indicators because they represent the three most common ways for residents to interact with their neighbours in urban China (Wang et al., 2016; Wu and Logan, 2016). The survey asked migrant respondents how often they had neighbourly interaction with local Shanghai residents whilst native Shanghai respondents were asked how often they interacted with non-local residents. The three neighbouring questions were asked in the following form and respondents could choose from four answers: (1) frequently (2) sometimes (3) seldom (4) never.

With regards to how out-group is conceptualised, migrant residents were asked about their relationships with local Shanghai residents (*bendiren*) whilst local Shanghai residents were asked about their relationships with non-local residents (*waidiren*). This definition of out-group was chosen because it is currently one of the most commonly used definitions of one's own social identity.

Independent neighbourhood level variables

Table 9.2 presents the list of independent variables. This study uses three contextual variables at the neighbourhood level, which were all obtained from the respective local residential committees (*juweihui*), the de-facto government institution at the grassroots level. The percentage of migrant residents at the *juweihui* level is included as a continuous variable to represent migrant density. The housing type of the neighbourhood is included as a categorical variable in order to account for the differing lifestyles of residents living in housing types. Due to the lack of official data on poverty, to measure neighbourhood poverty, this study follows Wu et al.'s (2010) approach and uses the number of Minimum Living Standard Support (MLSS) recipients within the *juweihui*. The MLSS recipient indicator is included as a continuous variable and was standardised to account for the variance of population between neighbourhoods. Additionally, considering that non-local hukou holders do not receive any MLSS support from the government, we included the neighbourhood type of urban villages to reflect migrant poverty to a certain degree. The study of Wu et al. (2010: 140) also shows that urban villages belong to the lowest social stratum in terms of poverty and primarily consist of rural migrants.

Table 9.2 Independent variables

		Total	%
Gender	male	806	56.76
	female	614	43.24
Education level	Elementary or below	148	10.42
	Junior secondary	489	34.44
	Senior secondary	349	24.58
	College or technical school	243	17.11
	Undergraduate or Master	191	13.45
Income	0–3000 Yuan	155	10.92
	3000–5000 Yuan	312	21.97
	5000–7000 Yuan	352	24.79
	7000–10,000 Yuan	350	24.65

		Total	%
	>10,000 Yuan	251	17.68
Tenure	tenant	356	25.16
	owner	1059	74.84
Years of residence	mean	18.1	
Household size (number of family members)	minimum	0	
	maximum	6	
	mean	1.75	
Percentage of migrant residents	minimum	0	
	maximum	76	
	mean	26.93	
Area poverty (number of MLSS recipients)	minimum	0	
	maximum	80	
	mean	12.35	
Housing type	Courtyard housing	156	11.02
	Work unit	411	29.03
	Urban villages	205	14.48
	Relocation housing	252	17.80
	Commodity housing	392	27.68

Independent individual level variables

The analysis includes demographic and socio-economic variables such as age, income and number of family members to account for individual effects on neighbouring. All independent variables have been standardised. Furthermore, factors of neighbourhood life such as length of residency as well as tenure were used as control variables as these indicators have been found to be significant determinants of neighbouring (Li et al., 2012). This study used hukou status as a key individual determinant in order to explore the difference between locals and non-locals in terms of their propensity to interact with out-group neighbours. There are four categories of hukou status: local non-agricultural (native urban), local agricultural (native rural), non-local agricultural (rural migrant) and non-local non-agricultural (urban migrant). This study also included several interaction terms to account for the various differences between native residents, urban and rural migrants. The interaction terms are income, area poverty and migrant density.

Statistical method

This study used a mixed effects linear regression approach, also known as multilevel modelling, to investigate the determinants of the neighbourly relations between migrants and locals. The reason for adopting a multilevel approach rather than an Ordinary Least Squares (OLS) model is because a standard OLS model cannot take into account the hierarchical structure of the survey data (Gelman and Hill, 2006; Raudenbush and Bryk, 2002). Although the survey data was collected using a random selection process, it is nevertheless impossible to disregard the 'natural' hierarchy of the data (Raudenbush and Bryk, 2002). An example of a natural hierarchy could be that children (Level 1) are nested within parents (Level 2) and therefore the probability that children from the same parents resemble each other more than compared to a random selection is considerably higher. In the case of this study residents (Level 1) are nested within the neighbourhood (Level 2) and a strong possibility existed that certain population subgroups would congregate in specific neighbourhoods. In terms of income for instance, considering that residential segregation is based on socio-economic status in Shanghai (Li and Wu, 2008; Shen and Xiao, 2020) it is therefore assumed that the similarity of the income level is generally higher between residents living in the same residential neighbourhood as compared to a randomly sampled individual. A standard regression model would be unable to take into account this hierarchical structure and therefore produce over-optimistic results and large standard errors (Gelman and Hill, 2006). For this study it was therefore necessary to apply a mixed effects model in order to allow the intercept to vary across different neighbourhoods. The benefits of using the multilevel model approach have also been approved by most studies researching intergroup relations at the neighbourhood level (Gundelach and Freitag, 2014; Laurence, 2011; Letki, 2008; Putnam, 2007; Secor and O'Loughlin, 2005).

The mixed effect linear model (or random intercept model) used in this study can be expressed as the following equation:

$$y_{ij} = \beta_{0ij} + \beta_1 x_{1ij} + \beta_2 x_{2j} + u_j + e_{ij} \tag{9.1}$$

where y represents the dependent variable such as the frequency of intergroup neighbouring activities. *i* denotes that variable varies between individual (Level 1) and *j* denotes that variable varies between *juweihui* (Level 2) and is constant for all respondents within a given neighbourhood. $\beta_1 x_{1ij}$ are individual predictor variables such as age or income. $\beta_2 x_{2j}$ are neighbourhood level predictors such as the poverty rate of the area or the neighbourhood housing type. u_j is the Level-2 residual and e_{ij} is the Level-1 residual.

Results

Table 9.3 shows the mixed effects linear regression results of the intergroup neighbouring activities between migrants and locals in Shanghai. Before analysing the mixed effects linear model, the study sought to find out whether it is statistically meaningful to adopt a multilevel model approach. We carried out a likelihood ratio test to compare the estimation between a conventional OLS model and a mixed effect model (Gelman and Hill, 2006). The likelihood ratio test shows that neighbourhood variance is significant ($p<0.001$) and that allowing for the neighbourhood level intercept to vary can significantly improve the estimations compared to the OLS model. Consequently this result justifies the usage of a mixed effects model. Moreover, it is useful to note that neighbourhood variables contribute about 28.9% to the variance of the overall model.

Table 9.3 Determinants of intergroup neighbouring (N=1405, weighted)

		B	S.E.
Constant		3.940***	0.601
Neighbourhood level			
Area poverty		-0.322	0.308
Neighbourhood type	Courtyard housing	0.012	1.057
	Work-unit	1.028	0.582
	Relocation housing	0.877	0.662
	Commodity housing	1.018	0.667
	Urban villages (reference)		
Migrant concentration		1.187**	0.376
Individual level			
Age		0.234**	0.076
Hukou status	Rural local hukou	0.234	0.496
	Urban migrant hukou	1.663***	0.225
	Rural migrant hukou	1.401***	0.310
	Urban local hukou (reference)		
Education level		0.073	0.054
Income		0.089	0.083
Tenure	Tenant	-0.025	0.138
	Owner (reference)		
Years of residence		-0.093	0.084
Household size	Number of family members	0.128*	0.061
Interaction terms			
Hukou and income	Local rural	-0.207	0.284
	Urban migrant	-0.347*	0.151
	Rural migrant	-0.166	0.124
	Local urban hukou (reference)		

(Continued)

Table 9.3 (Continued)

		B	S.E.
Hukou and migrant concentration	Local rural	-0.383	0.339
	Urban migrant	-0.737***	0.202
	Rural migrant	-0.422*	0.188
	Local urban hukou (reference)		
Hukou and area poverty	Local rural	0.121	0.110
	Urban migrant	0.256	0.316
	Rural migrant	0.544*	0.242
	Local urban hukou (reference)		
Within area variance		2.878	0.201
Between area variance		1.171	0.259

Notes: * $p<0.05$; ** $p<0.01$; *** $p<0.001$; significance $p<0.001$

With regards to the neighbourhood level determinants, migrant concentration is the only significant factor. More importantly, the level of significance ($p<0.01$) suggests that the presence of migrant residents is an important indicator of intergroup neighbouring whereby higher numbers of migrant residents lead to higher levels of intergroup neighbouring activities. This finding confirms the second hypothesis (H2) and implies that higher presence of migrant residents increases intergroup neighbourly relations. Furthermore, the interaction between migrant density and hukou status reveals that, in comparison to urban local residents, migrant residents living in areas with a high migrant presence are significantly less likely to interact with their native residents. This outcome is of little surprise when considering that in areas with a strong migrant concentration, migrant residents can choose to completely rely on their fellow in-group neighbours rather than reach out to their native residents. This may be particularly true for migrants living in migrant enclaves where the extreme concentration of migrants and heavy dependence on in-group social ties may prevent any interactions with local residents. Earlier research into urban villages also confirms that in migrant enclaves, local and migrant residents tend to have a very distant relationship (Chung, 2010). With respect to local residents, the effect of higher migrant presence also increases the likelihood of native Shanghai residents' neighbouring frequency with migrant residents. This suggests that having more migrant neighbours does not lead to local residents retracting from the neighbourhood, thus rejecting both the conflict theory (Vervoort, 2012) and the constrict theory (Putnam, 2007). Most likely having more migrants in the neighbourhood increases the exposure rate to migrant neighbours, which in turn helps overcome existing stigmas and perceived social gap of native urban Shanghai residents. However, again it is important to note that an extreme concentration of migrant residents may lead to the reverse.

Although the area poverty variable was not significant, the interaction term of area poverty and hukou status reveals that rural migrants living in poorer neighbourhoods are actually more likely to have neighbourly interactions with their native neighbours ($p<0.05$). This finding confirms hypothesis 3 (H3) and contradicts existing studies on multi-ethnic societies where area deprivation is often found to be detrimental to interethnic neighbourly relations (Laurence, 2011; Letki, 2008). However, this is not surprising when considering that rural migrants who live in poorer areas are more marginalised and therefore reliant on local support networks. Recent evidence also shows that rural migrants are already employed before moving their home. This indicates that no job competition exists between migrants and locals in poor neighbourhoods (Li and Zhu, 2014). Additionally, neighbourhood type is found to be not significantly associated with intergroup neighbouring, which is contrary to earlier research (Forrest and Yip, 2007; Wu and He, 2005). One potential reason might be because housing type can influence whether residents wish to engage in neighbourly activities but cannot affect *whom* residents are willing to interact with. For instance, native urban residents in traditional courtyards may be more dependent on local ties but they can choose to mostly interact with other native urban neighbours and not with migrant neighbours.

With regards to individual level predictors, hukou status is a highly significant determinant of intergroup neighbouring whereby both urban ($p<0.001$) and rural migrants ($p<0.001$) are more likely to have bridging neighbourly interactions. The finding confirms hypothesis 1 (H1) and implies that migrant residents, regardless of their hukou status, generally tend to reach out to native residents as a means to settle down in a foreign environment. Neighbouring in this sense serves an important purpose of creating social ties in the host society. Aside from economic benefits of out-group relationships, migrants may also seek intergroup relations as means to feel more included in the local community. In comparison, native residents do not have this need to create social ties with migrants, as they are already members of the host society. Additionally, the model shows that older residents are more likely to interact with out-group neighbours, which may be because older residents tend to spend more time locally. Having more household members is also positively related to intergroup neighbouring ($p<0.05$) since more family members may also increase the chance encounter of a household with other neighbours. Finally, the interaction term between income and hukou status reveals that more affluent urban migrants are less likely to have neighbourly interactions with native residents. The reason may be because urban migrants tend to be from the middle class and have a less localised social network. Middle class urban migrants may also prefer privacy in their living environment rather than socialising with neighbours.

Conclusion

Neighbourhood social interaction has gained significant attention in recent years in urban China due to the need to socially integrate increasingly larger numbers of migrant citizens. However, despite the growing focus on intergroup relations, measuring the quality of such relations in urban China remains challenging. Problems such as air pollution or climate change, which have tangible indicators, can be analysed through new data collection methods. However, social relations are often based on sentiments and are much harder to measure by data collection methods that rely on new technologies. Furthermore, in China, inability to access large-scale surveys carried out by the Chinese government only renders research into social issues all the more challenging. Under such a 'data scarce' context, collecting first hand data through household surveys becomes an important means to explore emerging social problems.

This chapter contributed to a better understanding of the social integration of domestic migrants in urban China through investigating the neighbourly relationship between rural migrants and native urban residents in Shanghai. China, like many other fast developing countries such as countries in sub-Saharan Africa (Englund, 2002; Erman, 1998; Gugler, 2002), is struggling to integrate its growing numbers of domestic migrants. In urban China, rural migrants often feel discriminated against by native urban residents and feel socially excluded from the urban society. Existing research suggests that better intergroup relations with indigenous residents can help migrants feel better integrated and accepted into the host society (Letki, 2008; Nielsen and Smyth, 2011). However, what affects intergroup relations in urban China remains relatively unknown. This study contributes to this lacuna and found that the stigmatisation of migrants and lack of out-group residents reduces the likelihood for migrant and local residents to form intergroup relations. On the other hand, we found that the need for informal support is likely to stimulate more intergroup neighbourly relations. The analysis results show that migrant residents are more likely to interact with their native neighbours than the other way around. This is mainly due to the fact that migrants are in more need of a diverse social network in order to survive and advance further in the host society, which also explains why migrants living in poorer neighbourhoods tend to have more intergroup ties with their native neighbours. Considering the tremendous institutional limitations placed on rural migrants due to the hukou system, native residents form an important source of support and information (Liu et al., 2013; Yue et al., 2013). Migrants are driven by the need to survive and to socially settle down in a foreign society through establishing contacts with locals whereas for indigenous hukou holders not possessing any social ties with migrants will not lead to any losses on their side since they are already full members of the urban society. To put it in simple terms, there are no tangible gains for local residents by having any social ties with migrant residents. Consequently, compared to migrants, the reasons why certain indigenous residents interact with their migrant neighbours may be more of a benevolent and voluntary nature, which is why considerably fewer locals are practising it. Finally, the analysis results indicate that neighbourhoods with a more balanced share of both migrant and local residents appear to be most conducive to intergroup neighbouring. The reason for this could be that having more out-group members in the neighbourhood simply increases the chances for residents to encounter each other and therefore facilitate interaction. Moreover, with migrant residents being a significant part of the neighbourhood, this also suggests that their cooperation is needed in solving local issues.

With respect to the policy implications of this study, our findings suggest that the hukou registration system has significantly reduced migrant residents' ability to integrate into the city. The hukou system has resulted in the stigmatisation and discrimination of rural residents which in turn inhibits their ability to form social ties with local residents. Removing the hukou system or at the very least removing barriers to welfare entitlements for rural migrants is a crucial step in aiding their integration into the city. In terms of implications for urban planning, currently China is adopting a demolition based regeneration approach which particularly targets older neighbourhoods with higher levels of migrant residents. Migrant enclaves such as urban villages are also targeted for demolition since they are regarded as sources of crime and poverty. Our findings however, show that older neighbourhoods with many migrant residents are an important platform for them to form positive relationships with native residents. Instead of demolition, neighbourhoods with higher migrant residents should be preserved and used as the starting point for many integration and welfare support schemes targeting migrant residents.

By focusing on the issue of intergroup relations, this chapter shows that traditional methods of data collection through household surveys still play a vital role in studying social issues especially in contexts where data is scarce. New data collection methods are effective in measuring and visualising complex urban problems that had been difficult to study only a few decades ago. However, they are relatively limited when it comes to analysing intangible concepts such as intergroup tolerance and social integration. Often such issues still require more traditional data collection methods such as questionnaire surveys. Furthermore, this study has also shown the shortcomings and challenges of carrying out household surveys. Certain population groups are often harder to access for various reasons and the resulting survey sample may not be as representative. Adopting a robust sampling technique and budgeting for unexpected expenses are helpful ways to mitigate these shortcomings. For our survey, we noticed that migrant respondents were underrepresented due to their long working hours and irregular work patterns. To address this problem, we carried out an additional 100 questionnaires specifically targeting migrant respondents and made sure to adjust to their working hours. Nevertheless, for this study, a household survey has been an indispensable source of information to understand the social integration and intergroup relations of rural migrants in urban China. Finally, despite the importance of household surveys, there is also great potential to introduce new data collection methods to better understand human behaviour and attitude towards other social groups. Recent works have already shown some promising advances in using modern technology to study traditionally sociological issues. For instance, recent studies have made great use of mobile phone data and GPS tracking to research the social segregation in Hong Kong (Li and Wang, 2017) and the social equity of access to urban parks in Shanghai (Xiao et al., 2019). The downside of such forms of data is that they cannot be connected to the respondent's socio-economic profile (e.g. employment, income, etc.). For instance, data collected from mobile phone data can help reveal the travel pattern and usage of open space of respondents (Xiao et al., 2019) but one does not know the demographic and socio-economic status of respondents.

The lack of such information can in turn limit an investigation into how, for instance, social class can affect one's residential preferences or places to visit during one's leisure time. One way to address this could be for instance to combine traditional household surveys with modern technologies. The advantage of combining both methods is that it would be possible to connect a respondent's demographic profile and socio-economic status with specific indicators (e.g. travel pattern). In the case of neighbouring studies for instance, it would be very interesting to find out where residents spend their leisure time, as a way to gauge their involvement in the local area, and how this behaviour might be related to their socio-economic status. However, in a context of cyber security and privacy, endeavours to combine survey and modern technologies must proceed with great caution and ensure that they will in no way harm the privacy and safety of respondents.

References

Alesina, A. and La Ferrara, E. (2002) 'Who trusts others?', *Journal of Public Economics*, 85 (2): 207–34.

Allport, G.W. (1954) *The Nature of Prejudice*. Cambridge, MA: Perseus.

Buonfino, A. and Hilder, P. (2006) *Neighbouring in Contemporary Britain: A Think-piece for the Joseph Rowntree Foundation Housing and Neighbourhoods Committee*. York: Joseph Rowntree Foundation. Available from https://www.jrf.org.uk/report/neighbouring-contemporary-britain-think-piece-joseph-rowntree-foundation-housing-and (accessed January 8, 2021).

Chen, X., Stanton, B., Kaljee, L., Fang, X., Xiong, Q., Lin, D., Zhang, L. and Li, X. (2011) 'Social stigma, social capital reconstruction, and rural migrants in urban China: A population health perspective', *Human Organization*, 70 (1): 22–32.

Chung, H. (2010) 'Building an image of villages-in-the-city: A clarification of China's distinct urban spaces', *International Journal of Urban and Regional Research*, 34 (2): 421–37.

Englund, H. (2002) 'The village in the city, the city in the village: Migrants in Lilongwe', *Journal of Southern African Studies*, 28 (1): 137–54.

Erman, T. (1998) 'Becoming "urban" or remaining "rural": The views of Turkish rural-to-urban migrants on the "integration" question', *International Journal of Middle East Studies*, 30 (4): 541–61.

Forrest, R. and Yip, N.-M. (2007) 'Neighbourhood and neighbouring in contemporary Guangzhou', *Journal of Contemporary China*, 16 (50): 47–64.

Gelman, A. and Hill, J. (2006) *Data Analysis Using Regression and Multilevel/Hierarchical Models*. Cambridge: Cambridge University Press.

Guest, AM. and Wierzbicki, S.K. (1999) 'Social ties at the neighborhood level: Two decades of GSS evidence', *Urban Affairs Review*, 35 (1): 92–111.

Guest, A.M., Kubrin, C.E. and Cover, J.K. (2008) 'Heterogeneity and harmony: Neighbouring relationships among whites in ethnically diverse neighbourhoods in Seattle', *Urban Studies*, 45 (3): 501–26.

Gugler, J. (2002) 'The son of the hawk does not remain abroad: The urban–rural connection in Africa', *African Studies Review*, 45 (1): 21–41.

Gundelach, B. and Freitag, M. (2014) 'Neighbourhood diversity and social trust: An empirical analysis of interethnic contact and group-specific effects', *Urban Studies*, 51 (6): 1236–56.

Hazelzet, A. and Wissink, B. (2012) 'Neighborhoods, social networks, and trust in post-reform China: The case of Guangzhou', *Urban Geography*, 33 (2): 204–20.
Huang, X., Liu, Y., Xue, D., Li, Z. and Shi, Z. (2018) 'The effects of social ties on rural-urban migrants' intention to settle in cities in China', *Cities*, 83: 203–12.
Ihlanfeldt, K.R. and Scafidi, B.P. (2002) 'The neighbourhood contact hypothesis: Evidence from the multicity study of urban inequality', *Urban Studies*, 39 (4): 619–41.
Laurence, J. (2011) 'The effect of ethnic diversity and community disadvantage on social cohesion: A multi-level analysis of social capital and interethnic relations in UK communities', *European Sociological Review*, 27 (1): 70–89.
Laurence, J. and Bentley, L. (2016) 'Does ethnic diversity have a negative effect on attitudes towards the community? A longitudinal analysis of the causal claims within the ethnic diversity and social cohesion debate', *European Sociological Review*, 32 (1): 54–67.
Letki, N. (2008) 'Does diversity erode social cohesion? Social capital and race in British neighbourhoods', *Political Studies*, 56 (1): 99–126.
Li, F. and Wang, D. (2017) 'Measuring urban segregation based on individuals' daily activity patterns: A multidimensional approach', *Environment and Planning A: Economy and Space*, 49 (2): 467–86.
Li, S., Zhu, Y. and Li, L. (2012) 'Neighborhood type, gatedness, and residential experiences in Chinese cities: A study of Guangzhou,' *Urban Geography*, 33 (2): 237–55.
Li, S.-M. and Zhu, Y. (2014) 'Residential mobility within Guangzhou city, China, 1990–2010: Local residents versus migrants', *Eurasian Geography and Economics*, (4): 313–32.
Li, Z. and Wu, F. (2008) 'Tenure-based residential segregation in post-reform Chinese cities: A case study of Shanghai', *Transactions of the Institute of British Geographers*, 33 (3): 404–19.
Li, Z. and Wu, F. (2013) 'Residential satisfaction in China's informal settlements: A case study of Beijing, Shanghai, and Guangzhou', *Urban Geography*, 34 (7): 923–49.
Liu, Y., Zhang, F., Liu, Y., Li, Z. and Wu, F. (2017a) 'The effect of neighbourhood social ties on migrants' subjective wellbeing in Chinese cities', *Habitat International*, 66: 86–94.
Liu, Y., Zhang, F., Wu, F., Liu, Y. and Li, Z. (2017b) 'The subjective wellbeing of migrants in Guangzhou, China: The impacts of the social and physical environment', *Cities*, 60: 333–42.
Liu, Z., Wang, Y. and Tao, R. (2013) 'Social capital and migrant housing experiences in urban China: A structural equation modeling analysis', *Housing Studies*, 28 (8): 1155–74.
Moch, L.P. (2003) *Moving Europeans – Migration in Western Europe since 1650*. Bloomington, IN: Indiana University Press.
NBS (National Bureau of Statistics) (2017) *China Statistical Yearbook 2017*. Beijing: China Statistical Press.
Nielsen, I. and Smyth, R. (2011) 'The contact hypothesis in urban China: The perspective of minority-status migrant workers', *Journal of Urban Affairs*, 33 (4): 469–81.
Nielsen, I., Nyland, C., Smyth, R., Zhang, M. and Zhu, C.J. (2006) 'Effects of intergroup contact on attitudes of Chinese urban residents to migrant workers', *Urban Studies*, 43 (3): 475–90.
Pettigrew, T.F. (1998) 'Intergroup contact theory', *Annual Review of Psychology*, 49 (1): 65–85.
Pow, C.-P. (2007) 'Securing the "civilised" enclaves: Gated communities and the moral geographies of exclusion in (post-)socialist Shanghai', *Urban Studies*, 44 (8): 1539–58.
Putnam, R.D. (2007) 'E pluribus unum: Diversity and community in the twenty-first century' (The 2006 Johan Skytte prize lecture). *Scandinavian Political Studies* 30 (2): 137–74.

Raudenbush, S.W. and Bryk, A.S. (2002) *Hierarchical Linear Models: Applications and Data Analysis Methods*. London: SAGE.
Secor, A.J. and O'Loughlin, J. (2005) 'Social and political trust in Istanbul and Moscow: A comparative analysis of individual and neighbourhood effects', *Transactions of the Institute of British Geographers*, 30 (1): 66–82.
Shanghai Municipal People's Government (2014) *2014 Shanghai Statistical Yearbook*. http://tjj.sh.gov.cn/tjnj/zgsh/tjnj2014en.html (accessed January 8, 2021).
Shen, J. and Xiao, Y. (2020) 'Emerging divided cities in China: Socioeconomic segregation in Shanghai, 2000–2010', *Urban Studies*, 57 (6): 1338–56.
Stolle, D., Soroka, S. and Johnston, R. (2008) 'When does diversity erode trust? Neighborhood diversity, interpersonal trust and the mediating effect of social interactions', *Political Studies*, 56 (1): 57–75.
Twigg, L., Taylor, J. and Mohan, J. (2010) 'Diversity or disadvantage? Putnam, Goodhart, ethnic heterogeneity, and collective efficacy', *Environment and Planning A*, 42 (6): 1421–38.
Vervoort, M. (2012) 'Ethnic concentration in the neighbourhood and ethnic minorities' social integration weak and strong social ties examined', *Urban Studies*, 49 (4): 897–915.
Wang, Z., Zhang, F. and Wu, F. (2016) 'Intergroup neighbouring in urban China: Implications for the social integration of migrants', *Urban Studies*, 53 (4): 651–68.
Wang, Z., Zhang, F. and Wu, F. (2017a) 'Affective neighbourly relations between migrant and local residents in Shanghai', *Urban Geography*, 38 (8): 1182–202.
Wang, Z., Zhang, F. and Wu, F. (2017b) 'Neighbourhood cohesion under the influx of migrants in Shanghai', *Environment and Planning A*, 49 (2): 407–25.
Wang, Z., Zhang, F. and Wu, F., (2020) 'The contribution of intergroup neighbouring to community participation: Evidence from Shanghai', *Urban Studies*, 57 (6): 1224–1242.
Wu, F. and He, S. (2005) 'Changes in traditional urban areas and impacts of urban redevelopment: A case study of three neighbourhoods in Nanjing, China', *Tijdschrift voor Economische en Sociale Geografie*, 96 (1): 75–95.
Wu, F. and Logan, J. (2016) 'Do rural migrants "float" in urban China? Neighbouring and neighbourhood sentiment in Beijing', *Urban Studies*, 53 (14): 2973–90.
Wu, F., He, S. and Webster, C. (2010) 'Path dependency and the neighbourhood effect: Urban poverty in impoverished neighbourhoods in Chinese cities', *Environment and Planning A*, 42 (1): 134–52.
Wu, R., Huang, X., Li, Z., Liu, Y. and Liu, Y. (2019) 'Deciphering the meaning and mechanism of migrants' and locals' neighborhood attachment in Chinese cities: Evidence from Guangzhou,' *Cities*, 85: 187–95.
Xiao, Y., Wang, D. and Fang, J. (2019) 'Exploring the disparities in park access through mobile phone data: Evidence from Shanghai, China', *Landscape and Urban Planning*, 181: 80–91.
Yue, Z., Li, S., Jin, X. and Feldman, M.W. (2013) 'The role of social networks in the integration of Chinese rural-urban migrants: A migrant-resident tie perspective', *Urban Studies*, 50 (9): 1704–23.
Zhu, Y., Breitung, W. and Li, S. (2012) 'The changing meaning of neighbourhood attachment in Chinese commodity housing estates: Evidence from Guangzhou', *Urban Studies*, 49 (11): 2439–57.

10

CONCLUSION

BEYOND ANALYTICS: PLANNING-LED OR PLANNING-LAG?

Cecilia Wong

This concluding chapter was written amidst the global outbreak of Coronavirus (COVID-19). During the homebound lockdown days, there were all sorts of data, models, images, news and rumours of the virus from medical professionals, academic communities, government departments, and social media. This epitomises the explosion of data in different formats and channels. The promise of developing a 'world-beating' track and trace app in the UK to provide real-time data to monitor the spread of the virus ended up with the embarrassing conclusion that 'a human being on the end of a phone is a far better option' (Cellan-Jones, 2020). There are also controversies over the counting and reporting of death and infection numbers (Conway, 2020). This does verify many of our concerns over the weaknesses and inadequacy of our data infrastructure in a time when we urgently require prompt and reliable data to monitor the deadly virus.

Against this canvas of uncertainties and desperation, this book provides a timely and pertinent contribution to highlight the challenge of developing and applying quantitative methods and techniques to understand and manage planning problems in a brave and grave new world. Despite starting off the analysis from different perspectives and different spatial contexts, the opportunities and challenges identified by the authors are unexpectedly similar and their arguments often resonate with each other. Threads of their arguments are thus brought together in this concluding chapter to help develop the agenda for future research and practice.

The mismatch between urban issues, planning systems and data analysis

Since there has been an increasing emphasis on integrating different policy concerns within coherent spatial frameworks and initiatives (de Roo and Silva, 2010; Wong and Watkins, 2009), the traditional sectoral approach to examining techniques and methods (e.g. Bracken, 1981; Field and MacGregor, 1987; Roberts, 1974) no longer provides a meaningful way forward. Besides developing data-hungry grand statistical models to capture the complexity of spatial relationships, quantitative techniques and methods in spatial planning have become both analytical and communicative instruments. The purpose is that different methods and techniques should be flexibly combined to tackle specific policy analytical needs within specific spatial contexts.

With uncertainties brought by natural disaster, infectious disease, social unrest and economic turbulence at this juncture of time, Silva et al. rightly argue in Chapter 2 that urban problems are complex and unpredictable and require planning systems that are adaptable and able to cope with the dynamics of change. Their debate is firmly embedded in the complexity theoretical perspective that cities are dynamic intertwining systems and adaptive planning is thus required to manage urban development trajectories. Given that planning systems are mostly static, with a lengthy duration of data collection and analysis, this makes them rather unsuited to manage complex urban issues within a smart city environment where a data science approach is desired.

Planning is applied at a variety of spatial scales, ranging from strategy development at the regional and sub-regional or city-regional scale to master planning of local neighbourhoods, as well as addressing cross-boundary issues. The tension between the administrative geography of a planning system as a government institution and the functional spatial scale of real-life phenomena has been a long-standing conundrum. The criss-crossing of multi-scalar policy layers over different temporal dimensions has created both complexity and challenges to measurement. Wong's discussion of indicators as monitoring instruments of planning in Chapter 3 highlights this spatial-temporal methodological challenge, which is echoed by other contributors in the book (e.g. Dawkins et al. and Silva et al.). The advance of new analytical techniques for different sources of big data has offered promising prospects of overcoming these challenges. The discussion of the development of WUDAPT in Chapter 7 by Ren et al. shows how the application of Landsat satellite images is used to derive robust data on changes in different categories of urban forms and functions to support climate change. Likewise, multi-scalar regional accessibility assessment of high-speed rail development in the Yangtze River Delta was established by Lei Wang through painstakingly integrating and aggregating different datasets from the grid level into larger administrative units (see Chapter 8).

Another key characteristic of planning is its 'wicked' nature (Rae and Wong, 2012) nature of being embedded in a dynamic social context interacting with a wide variety of actors and practices. This poses challenges to developing robust measurements of planning phenomena (Rae and Wong, 2012; Wong, 2006). To Rittel and Webber (1973), only by laying out alternative understandings of problems, competing interests, priorities and constraints is it then possible to apply more formal analytical tools to understand a specific problem scenario. Interestingly, Kitchin (2014) points out that big data is socially situated and the neglect of its socio-political nature will decontextualise its usage and analysis from the wider set of social, economic, cultural and environmental relations and the everyday experience of urban process and change. Being sensitive to the context of planning and the nature of big data highlights another important challenge when applying urban analytics to address planning issues.

The intersection of technology and governance

While recognising that planning issues and big data are both complex and socially situated, Smith in Chapter 5 further points out the existence of digital divides among different groups of users as well as institutional divides over the control of data access. Being context-sensitive thus brings up the importance of understanding the intersection of technology and governance in planning. Scott (1998) comments that the ways governments tend to work is to skew decision-making towards centralised systems of control and favour expert knowledge to force particular ways of 'seeing' policy problems and potential solutions. This often results in the marginalisation of particular groups or perspectives.

The potential use of GIS and digital visualisation platforms to share knowledge across departments and to disseminate ideas in public nonetheless does not sit well with traditional centralised management and closed expert policy networks (Vonk et al., 2005; Wong et al., 2015). As highlighted by Silva et al. in Chapter 2, data visualisation techniques with detailed information from the rapidly expanded family of urban sensors used in simulation can provide better interfaces for policymakers and the public to understand the models. Complexities in planning issues require policymakers to question normative assumptions and stakeholders to rethink the principles of spatial organisation and activities. The challenge however lies in balancing the presentation of complex concepts in a form which is understandable by a range of actors. Smith in Chapter 5 helpfully illustrates the challenge of striking a balance between exploratory, multi-dimensional analysis and presentational, narrative-driven functionality through different design techniques such as interface metaphors, controlling levels of user interaction and creating data narratives.

Task-focussed analytics tools are found by Smith to be particularly relevant for urban planning contexts, whereas data journalism and storytelling can make data-driven visualisations more accessible for general audiences. The creation of multiple visualisations for different users is shared by Dawkins et al.'s discussion of the re-designing of the Dublin city dashboard through the use of themes, stories, tasks and tools as the designing architecture and bringing the potential of multimodal data interfaces of 3D environments in Chapter 6. They also remind us of the importance of Hansen's 'User Engineering Principles' (Hansen, 1971) by arguing that the requirements, skills and knowledge of the targeted users should be seriously considered when designing city dashboard systems. Unfortunately, many creators of city dashboards simply make assumptions rather than create meaningful engagement with their users.

Indeed, users can be the key drivers of data analytics through crowdsourcing information and community-based participation. In Chapter 4, Yudono has charted the progress made by raising public awareness of updating spatial data voluntarily on buildings, roads and other infrastructure through digital mapping to support government activities in disaster management in Indonesia. The delivery of information from citizens to government agencies has resulted in spatial visual images used as the negotiation medium to achieve planning and development consensus. This illustrates how citizen participation in a crowdsourced geographic information context has advanced demographic governance. The open-sourced development approach of the WUDAPT, as explained by Ren et al. in Chapter 7 also shows us the importance of global, bottom-up, self-organised effort to apply Landsat satellite images with machine learning to derive information about the form and function of cities as input data for climate change modelling.

The conundrums of data science

The advance of data science has provided us with promising potential to have new and sophisticated planning instruments, increasingly in real-time and with different types of visualisation functions. However, many authors in this book also highlight the limitations and challenges of these emerging technologies and data sources. We are still in the experimental stage of these emerging analytical techniques.

Transaction costs and marginal gains

The many advantages brought by the data rich environment have come with the challenge over how to collect, manage and integrate large and diverse datasets with unstructured formats (Silva et al., 2020) as well as dealing with the constraints of technical know-how, ethical considerations, the regulatory environment, and available resources (Kitchin, 2014). Much of 'big' data is not open sourced but held by the private sector or public organisations under strict privacy laws and data protection regulations. Another challenge is the heterogeneity of urban datasets with variability in data types and formats. All these constraints mean that the use of big data does come with major transaction costs of data mining and data integration as well as the compliance of different data access requirements and regulations. For example, Lei Wang found that the accuracy and reliability of different data sources involved in the assessment of high-speed rail accessibility vary widely and vary spatially, which required major effort to establish the accuracy and integrate the data. That is why the WUDAPT project, as explained by Ren et al., tries to develop a protocol and common practice to link different data for climate modelling. Although digital mapping in Indonesia is based on bottom-up community projects, as Yudono commented, its progress tends to be halted when the associated project budget to support the endeavour runs out.

Data rich, intelligence poor

With continuous improvements in the availability of urban data and analytical software, the assumption is that there will be a promising prospect of creating useful information and intelligence. However, as Shearmur (2015) points out, despite the 'big' size, the scope of big data is indeed very narrow as such data tends to be passively generated by users and is market-driven and does not include the attributes of individuals which would allow more nuanced analysis. Its automatic and spontaneous nature means that it is not structured or framed by concepts and definitions that are most relevant to address research needs and policymaking. Through the work of the Consumer Data Research Centre, Smith finds that whilst the exploratory mapping approach is highly effective at highlighting spatial variations in socio-economic built-environment data, there are difficulties in adding statistical analysis and customising the interface due to the long development time required. Despite their recognition of the utility and value of a city dashboard as a mode of communication and means of making sense of the city, Dawkins et al. expressed reservation over its function to produce knowledge about cities for urban planning and management. This is due to the caveat over issues of data access, quality, comprehensibility and scope, as well as a more fundamental reframing of the use of the urban dashboard from a critical GIS perspective. For many of us there is a sense of *deja vu* about this debate, as we rehearsed similar issues about the use of GIS in policymaking in the 1990s (e.g. Gilfoyle and Wong, 1998; Worrall, 1994).

Knowledge know-how and participation

The case study examples offered by our authors highlight the fact that, with the new technology and sheer amount of new data that is on stream, the trick to exploiting data science for popular and democratic application is to master the knowledge know-how. As shown by Yudono, the success of developing crowdsourced geographic information with residents in Indonesia was through the support of the Humanitarian OpenStreetMap Team and the Australia-Indonesia Facility for Disaster. Another example is the WUDAPT project that, as Ren et al. explained, has close association with the International Association for Urban Climate which includes experts such as architects, planners, climatologists and meteorologists globally. Dawkins et al. concede that while city dashboards are targeted for the masses, they tend to be used by a small group of advanced users with sufficiently high level of skills to use the tools. This brings up the pertinent issue by Smith of how data journalism can be made more meaningful and accessible to wider audiences by improving user relatability and more advanced narrative functionality of analytics and visualisation applications. This can only be achieved if the tools can strike a fine balance between the exploratory/multi-dimensional analysis for the more technical astute professional and the presentational/narrative-driven functionality targeted at the wider public.

Complementary rather than displacement

The authors in this book have consistently pointed out that the varied and incomplete coverage of big data can create bias and errors in measurement and analysis and there is an emerging consensus that a hybrid approach between the use of big data and traditional data sources together is the best way forward. The mix use of big data with other independent data sources is also recommended by the wider academic community (e.g. Batty, 2013; Glaeser et al., 2018; Shearmur, 2015). In this book, Dawkins et al. recommend that 3D geospatial technologies should be used reflexively and in conjunction with other forms of knowledge and data to provide a more situated view to serve the ends of individuals and communities. Both Ren et al. and Lei Wang show that big data such as remote sensing, landuse data and crowdsourced information works best alongside traditional census and administrative data. While GPS tracking via mobile phone can show individual movement, such data cannot replace traditional surveys to provide in-depth analysis of social relationships. As demonstrated by Zhen Wang, social surveys still hold their value in terms of analysing the socio-economic attributes of individuals and the dynamics of intergroup social relations. Despite the lack of consistency and the variability of big data, its experimental use can encourage communities to use the information as a collective way of 'seeing' to inform their own data infrastructure and engagement with policy debate (Wong et al., 2015). This is echoed by Silva et al. who suggest that big data and data analytics can triangulate with traditional data to address the medium- and long-term urban issues such as public participation, social segregation, regional connections and functional divisions. Such integration has proved to be successful in countries even with less advanced data infrastructure. As shown by Yudono, the National Mapping Agency of the Indonesian government have proactively integrated official spatial data with crowdsourced geographic information based on the OpenStreetMap platform for disaster planning.

Back to the future: planning-led or planning-lag?

By integrating, processing, analysing and visualising large amounts of both structured and unstructured data, increasingly real-time, from a unified geo-referenced data platform, big data opens up possibilities to examine urban questions that were previously found impenetrable (Glaeser et al., 2018; Pan et al., 2016). As discussed above, the development of smart city analytics is still at the early stage (Anejionu et al., 2019) and there is a need to be mindful of the limitations and potential errors. The challenges discussed by our authors interestingly resemble those we have previously encountered during the quantitative revolution in the 1970s and the use of GIS in the 1990s (Gilfoyle and Wong, 1998; Wong, 1998). Silva et al. helpfully remind us that the seven sins postulated by Douglass Lee (1973) – hypercomprehensiveness, grossness, wrongheadedness, hungriness, complicatedness, mechanicalness, and expensiveness – remain valid when shedding a critical light on urban computation and urban data analytics. The contributions in this book show that we have been too preoccupied to address the critical issues identified by Kitchin and McArdle (2017) in relation to epistemology, scope and access to data, data quality and ecological fallacies, usability and data literacy, appropriate application, and socially responsible design.

The advance of computation technology and data science can drive planners into a passive position; but if planners are not on top of their game, our planning intelligence and analysis will be data-driven rather than issue-driven. As Dawkins et al. found, while it is relatively straightforward to display a 3D model, it is much more complex when importing other geospatial data and adding visual analytics into these models to address more nuanced planning issues. More importantly, the application of such data instruments is in fact laden with politics and ideology and will project different urban futures and promote certain visions. From the previous experience with the planning application of GIS, it is problem-framing and policy knowledge that helps to extract relevant intelligence for policymaking (Gilfoyle and Wong, 1998; Worrall, 1994). Hence, it is important to make sure that data science should be deployed as an effective and powerful analytical instrument for urban planning, rather than driving planning decisions down a blind alley.

Another interesting observation is that the production of new forms of data does involve complex processes of data mining and integration, which in turn requires complex data regulation and management to be in place. Smart technology cannot be effectively applied without smart management and smart people. This will involve major institutional reforms of our planning systems involving a major re-think of how urban planning can be conducted in a way that is smart, dynamic and adaptable to change. We have also to work out how best to train a new generation of urban planners and managers who have a strong grip of policy issues as well as being technically astute to master the know-how to deploy these technologies for smart urban planning.

The emerging application of data science presents a doubled-edge sword to the planning field – will it be planning-led or planning-lag in the face of the digital and big data era? As Dawkins et al. point out, only time will tell whether smart data science such as city dashboards and 3D geospatial technologies will constitute disruptive innovations that radically transform the praxes of planning.

References

Anejionu, O.C.D., Thakuriah, P., McHugh, A., Sun, Y., McArthur, D., Mason, P. and Walpole, R. (2019) 'Spatial urban data system: A cloud-enabled big data infrastructure for social and economic urban analytics', *Future Generation Computer Systems*, 98: 456–73.

Batty, M. (2013) 'Big data, smart cities and city planning', *Dialogues in Human Geography*, 3 (3): 274–9.

Bracken, I. (1981) *Urban Planning Methods: Research and Policy Analysis*. London: Methuen.

Cellan-Jones, R. (2020) 'Coronavirus: What went wrong with the UK's contact tracing app?', *BBC News*, June 20. https://www.bbc.co.uk/news/technology-53114251 (accessed February 10, 2021).

Conway, E. (2020) 'Coronavirus: Dodgy data and double counting – the UK's testing data is a mess', *Sky News*, June 3. https://news.sky.com/story/coronavirus-how-did-the-uks-covid-19-testing-data-become-such-a-mess-11999797 (accessed February 10, 2021).

de Roo, G. and Silva, E.A. (2010) *A Planner's Encounter with Complexity*. Farnham: Ashgate.

Field, B. and MacGregor, B. (1987) *Forecasting Techniques for Urban and Regional Planning*. London: Hutchinson.

Gilfoyle, I. and Wong, C. (1998) 'Computer applications in planning: Twenty years' experience of Cheshire County Council', *Planning Practice and Research*, 13 (2): 191–7.

Glaeser, E.L., Kominers, S.D., Luca, M. and Naik, N. (2018) 'Big data and big cities: The promises and limitations of improved measures of urban life', *Economic Inquiry*, 56 (1): 114–37.

Kitchin, R. (2014) 'The real-time city? Big data and smart urbanism', *GeoJournal*, 79: 1–14.

Kitchin, R. and McArdle, G. (2017) 'Urban data and city dashboards: Six key issues', in R. Kitchin, G. McArdle and T. Lauriault (eds), *Data and the City*. London: Routledge. pp. 111–26.

Lee, D.B. (1973) 'Requiem for large-scale models', *Journal of the American Planning Association*, 39 (3): 163–78.

Pan, Y., Tian, Y., Liu, X., Gu, D. and Hua, G. (2016) 'Urban big data and the development of city intelligence', *Engineering*, 2: 171–8.

Rae, A. and Wong, C. (2012) 'Monitoring spatial planning policies: Towards an analytical, adaptive and spatial approach to a "wicked problem"', *Environment and Planning B: Planning and Design*, 39: 880–96.

Rittel, H. and Webber, M. (1973) 'Dilemmas in a general theory of planning', *Policy Sciences*, 4: 155–69.
Roberts, M. (1974) *An Introduction to Town Planning Techniques*. London: Hutchinson.
Scott, J. (1998) *Seeing Like a State: How Certain Schemes to Improve the Human Condition Have Failed*. New Haven, CT: Yale University Press.
Shearmur, R. (2015) 'Dazzled by data: Big data, the census and urban geography', *Urban Geography*, 36 (7): 965–8.
Silva, E.A., Liu, L., Kwon, H.R., Niu, H. and Chen, Y. (2020) 'Hard and soft data integration in geocomputation: Mixed methods for data collection and processing in urban planning', in S. Gertman and J. Stillwell (eds), *Handbook on Planning Support Science*. Cheltenham: Edward Elgar. pp. 37–55.
Vonk, G., Gertman, S. and Schot, P.P. (2005) 'Bottlenecks blocking widespread usage of planning support sytems', *Environment and Planning A*, 37: 909–24.
Wong, C. (1998) 'Old wine in a new bottle? Planning methods and techniques in the 1990s', *Planning Practice and Research*, 13 (3): 221–36.
Wong, C. (2006) *Quantitative Indicators for Urban and Regional Planning: The Interplay of Policy and Methods*. London: Routledge.
Wong, C. and Watkins, C. (2009) 'Conceptualising spatial planning outcomes: Towards an integrative measurement framework', *Town Planning Review*, 80 (4/5): 481–516.
Wong, C., Baker, M., Webb, B., Hincks, S. and Schulze Bäing, A. (2015) 'Mapping policies and programmes: The use of GIS to communicate spatial relationships in England', *Environment and Planning B: Planning and Design*, 42 (6): 1020–39.
Worrall, L. (1994) 'The role of GIS-based spatial analysis in strategic management in local government', *Computer, Environment and Urban Systems*, 18 (5): 323–32.

INDEX

Page numbers in *italics* refer to figures; page numbers in **bold** refer to tables.

FSC
www.fsc.org
MIX
Papier | Fördert gute Waldnutzung
FSC® C083411

Kolibri 360 GmbH
Ferdinand-Jühlke-Straße 7
99095 Erfurt, Deutschland
produktsicherheit@kolibri360.de